DATE DUE

AG 1 4 08			

DEMCO 38-296

The Left Hand of Creation

The
Left Hand
of Creation

The Origin and Evolution
of the Expanding Universe

JOHN D. BARROW

JOSEPH SILK

New York Oxford
OXFORD UNIVERSITY PRESS

Oxford University Press

Oxford   New York   Toronto
Delhi   Bombay   Calcutta   Madras   Karachi
Kuala Lumpur   Singapore   Hong Kong   Tokyo
Nairobi   Dar es Salaam   Cape Town
Melbourne   Auckland   Madrid

and associated companies in
Berlin   Ibadan

Library of Congress Cataloging-in-Publication Data
Barrow, John D., 1952–
The left hand of creation : the origin and evolution of the
expanding universe / John D. Barrow, Joseph Silk.
p.   cm.
Originally published: New York : Basic Books, © 1983.
Includes index.
ISBN 0–19–508675–9.
ISBN 0–19–508676–7 (pbk.)
1. Expanding universe.   I. Silk, Joseph, 1942–   II. Title.
QB991.E94B37   1994
523.1′8—dc20   93–11046

9 8 7 6 5 4 3 2 1

Printed in the United States of America

Contents

Preface
to the New Edition

We have taken the opportunity afforded by the publication of this new edition to update the text with a survey of the principal developments that have occurred in cosmology since this book first appeared in 1983. More details of these developments can be found highlighted as footnotes to the main text throughout the book. The guide to further reading has also been revised to include some of the profusion of books about cosmology that have appeared during the past decade.

Introduction

When the first edition of this book was written cosmology was entering a period of transition. Previously it had been the preserve of astronomers and mathematicians: astronomers sought tell-tale remnants of the Universe's past in the light from distant stars while mathematicians struggled with the complexities of Einstein's great theory of relativity, seeking to extract its predictions about the possible structure of the Universe. But in the 1980s cosmology entered a new and exciting phase from which it may never emerge. It became a branch of physics. New ideas about the nature and number of the most elementary particles of matter focussed interest upon the behaviour of matter and radiation at the highest possible energies and temperatures—higher than any that can be achieved on Earth.

In order to test those expectations physicists had to search out new environments which display extremes of temperatures and density. The most extreme are those provided by the early moments of our Universe. Physicists took their theories about the behaviour of matter at high temperatures and looked for things that would result from them if they were applied to the early history of the Universe. By such reconstructions, it is possible to test which ideas about the behaviour of matter at high temperature are coherent, which lead to observable consequences for astronomy today and which lead to a present-day Universe like the one that we see. Astronomers examined these new theories to see if they could shed light on some of the astronomical universe's most puzzling features. This new emphasis upon bringing together the study of the very smallest things in Nature—elementary particle physics—and the very largest—extragalactic astronomy—has led to an explosion of interest in cosmology at all levels. Popular books expounding its ideas and discoveries abound; the population of young researchers and students working on it has grown dramatically; funding bodies have reorganised their traditional subject categories in order to deal with the new synthesis of 'particle astrophysics' and the professional conference circuit is dominated by meetings and workshops dedicated to its subject matter. This has had a trickledown effect upon student teaching programmes where the study of physics with astronomy has proved a draw to bring teenagers into the wider study of physics. No less significant has been the growing interest of the media and the general public in the major observational discoveries of modern cosmology and what they have to teach us about humanity's place in the Universe.

This influx of new ideas about the behaviour of matter under the conditions existing during the early history of the Universe has been matched by experimenters and astronomical observers. Astronomers have conducted huge surveys of the Universe in different wave bands of light. For the first time they have produced three-dimensional maps of the sources of optical light in the Universe. These surveys have been made possible by developments in light-detector technology. Modern quantum electronic devices, like those used in video cameras, register more than half the light that falls upon them whereas traditional photographic film collects only a percent of that light. Modern detectors can determine the distance to a far away galaxy in fifteen minutes where photographic film would require many hours to gather sufficient light. This advance allows large numbers of distant galaxies to be observed in a manageable period of time, and a very large catalogue of objects can soon be compiled.

The advent of fast, inexpensive, interactive computers has revolutionised the way astronomers attempt to explain the observed structure of the Universe. Instead of working solely with simplified mathematical descriptions of a complicated big problem, like the development of galaxy clustering, they reenact the process with the aid of large computers. By telling the computer the laws of gravity and providing it with a starting pattern of objects, one can watch how they move and subsequently cluster. The results can be compared with the observed clustering patterns in order to check our theories about the clustering of galaxies and to understand the starting conditions that applied in our Universe. These simulated histories are continually being made larger and more realistic. Gas and dust, together with the complicated heating and cooling processes that accompany their presence, are also being included in order to investigate the complex processes whereby clustering masses turn into luminous galaxies.

What have been the highspots in cosmology during the last decade? We will highlight a number of key areas where important developments have taken place. Some have seen the expansion of theoretical ideas but most have been associated with new observational discoveries. While some of these observations have provided dramatic confirmation of prior theoretical

Introduction

expectations, others came up with new and unexpected discoveries.

The primordial furnace

During the first months of the Big Bang, the universe was hot, at a temperature exceeding a million degrees Kelvin. The state of matter was that of a plasma, mostly electrons and protons. The density of electrons and protons was sufficiently high that the accompanying radiation field was locked into intimate thermal contact with the matter. As the expansion proceeds and the density drops, the radiation field at first remains firmly tied to the electrons. Eventually, thermal contact breaks down. When this happens, the spectrum of the radiation need no longer be that of a perfect blackbody if there is any extraneous source of energy. Such energy injection might have come from the decays of elementary particles or from the formation and death of the first generation of massive stars.

The Cosmic Background Explorer (COBE) satellite, launched by NASA late in 1989, has measured the cosmic microwave background to have the spectrum of a blackbody with unprecedented precision. The blackbody temperature is 2.726 Kelvin. Any deviations are less than a few hundreths of a percent. The very early universe evidently behaved like a perfect furnace of heat radiation. This also tells us that during its later evolution, when the first stars formed and radiated, any injection of energy into the background can have made only the slightest addition to the cosmic microwave background. At optical and infrared wavelengths, astronomers are still searching for a diffuse extragalactic background signal from the epoch of galaxy formation.

Tell-tale fluctuations

A more direct measure of early cosmic structure has succeeded. Early on, the Universe was a primordial fireball, dominated by radiation, and the occasional particles of nuclear matter were unable to agglomerate effectively against the pressure of the radiation. Embedded in the radiation field, however, are the imprints of the seed fluctuations in the distribution of matter from which all present-day structure originated. These tiny density fluctuations only begin to intensify under the inexorable force of gravity once the Universe is a few thousand years old. Between then and now, fluctuations grew in strength, eventually to develop into galaxies, galaxy clusters, and even larger structures. The Universe expanded by a factor of 100,000 or so in that period, and so the first seed fluctuations must have had an amplitude of one part in a 100,000 and should still be visible as hot (or cold) spots in the microwave sky. After two decades of searching, an experiment finally found them. In 1992, the COBE satellite reported the detection of temperature fluctuations in the microwave background near the expected level of 1 part in 100,000. These fluctuations are the elusive link between present-day structures and events close to the beginning of the Universe.

Interestingly, on a scale exceeding 10 degrees on the sky over which these fluctuations are detected (the full moon covers about half a degree on the sky), there has been insufficient time for any physical processes to have modified them since they were initially laid down. Nor is there any direct trace of them in the large-scale galaxy distribution. This implies that the largest measured inhomogeneities in the matter of the Universe around us must derive from smaller seed fluctuations on angular scales of degrees, rather than tens of degrees, in the microwave background.

Large-scale structure

On sufficiently large scales, over hundreds of megaparsecs, the galaxy distribution appears almost uniform. On smaller scales, structure emerges in the form of great clusters and superclusters of galaxies, and of apparent voids in the galaxy distribution. One can see glimmerings of such structure in two-dimensional maps of galaxies that are devoid of depth, which a terrestrial observer sees when viewing the sky. Our knowledge of structure has been deepened by the advent of three-dimensional maps, in which the galaxy redshift, related linearly to distance by Hubble's law of the expansion of the Universe, is used to provide a third dimension that probes depth in space. This mapping is slow and tedious, since spectra of many thousands of galaxies must be obtained. The results have provided a spectacular glimpse of structures that range up to tens of megaparsecs in size. Sheets, filaments and complex aggregates of galaxies are formed. Much of space is devoid of luminous galaxies. Whether dim galaxies are hiding in the voids is still not known.

Introduction

Moving streams of galaxies

Galaxies do not trace the expansion of the Universe perfectly. Just as waves are super-imposed on ocean tides, there are local to-and-fro motions that deviate from the universal expansion. These motions, driven by gravitational aggregations of galaxies, can be traced over tens of megaparsecs. Not only are galaxies moving in random directions within a cluster of galaxies, but the cluster itself, while participating in the over-all cosmic expansion, has a small, deviant velocity of its own. These flows amount to motions of hundreds of kilometers per second, and are powerful tracers of the dark matter distribution in the Universe. For it is known from the motions of galaxies within clusters, and of stars within galaxies, that at least 90 percent of the mass in the Universe is in some non-luminous form that must be generating these motions with its gravitational pull. These flows of galaxies tell us that the dark matter is clumpily distributed over tens of megaparsecs. Were it uniform over these scales, there would be no net force of gravitational acceleration acting on cluster-sized objects.

Intergalactic gas

The raw material from which galaxies were made is spread out in diffuse form throughout intergalactic space. Intergalactic hydrogen gas clouds were discovered by the absorption of light from distant quasars. The light absorption occurs at precisely the wavelength which marks the excitation of the hydrogen atom in its ground state. The expansion of the Universe creates a wavelength shift between the absorber and the observer. One can therefore infer the distance to the intervening clouds that are absorbing the radiation coming towards us from the distant quasars. Many such absorbing clouds are seen. These clouds are old, and formed when the Universe was just one or two tenths of its present age. The absorption spectra reveal that the clouds contain little pollution by stars, visible as heavy element features in the spectrum, and occupy large regions, corresponding to halos of galaxies. Such intergalactic clouds are plausible precursors to mature galaxies of stars.

On the scale of galaxy clusters, x-ray satellites have discovered an abundance of intergalactic gas. The gas is at a temperature of hundreds of millions of degrees Kelvin and glows in x-rays. Both in clusters and elsewhere, there is more diffuse intergalactic gas than there is mass in the form of ordinary stars. This is a testimony to the fact that star formation is highly inefficient, as indeed we see in our immediate vicinity throughout the Milky Way galaxy. Clues like this are pointers in our quest to develop a comprehensive theory of structure formation.

Galaxy formation

The key ingredients for forming galaxies are dark matter and diffuse hydrogen gas. The nature of the dark matter is elusive, but we can be reasonably sure that it consists of particles or massive objects that respond to gravity, but otherwise are weakly interacting, taking no part in electromagnetic or nuclear interactions. In this way, one can construct a galaxy with a bright disk or spheroid of stars, surrounded by a more extended, low density dark halo. About 90 percent of the total mass is in dark form, and dominates the mass of the galaxy. Dark matter controls the rate at which the luminous part of the galaxy can collapse, because it dictates the strength of the gravitational field.

Computer simulations have provided an attractive approach to modelling large-scale structure and give a broad brush view of galaxy formation. While the detailed microphysics of star formation is elusive, the gravitational pull of dark matter is simple to understand and straightforward to model. One can obtain maps of the galaxy distribution and computer-generated images of spiral or elliptical galaxies that are almost indistinguishable from the real thing. This is telling us that we are at least on the right track to decipher the origin of structure, even if we do not yet know the detailed composition of the dark matter or how star formation operates.

Primordial nucleosynthesis

Ten years ago, cosmological applications of high-energy physics were still relatively new. Attempts had been made to arrive at a single unified description of the strong and electroweak forces of Nature. These first 'Grand Unified Theories' (GUTs) offered an explanation of why the Universe was predominantly composed of matter rather than an equal mixture of matter and antimatter. Little has changed regarding that insight, but a by-product of it, the very slow decay of the proton, has yet to be observed despite a long search. Unfortunately, the most credible versions of these unified theories allow the proton to have a decay rate that is too slow for our detectors to see. Elsewhere, experimenters had a major success. The CERN collider produced Z bosons in great profusion. This enabled their subsequent decays to be monitored. One of the ways for the Z to decay is into pairs of neutrinos and their corresponding antiparticles (antineutrinos). Hence, the more varieties of neutrino there are for the Z to decay into, the faster it decays. So a careful study of Z decays can reveal how many varieties of neutrino there are which are light enough to be formed from the decay of a Z particle. The answer was three. This is the number expected in the standard model of particle physics. But it was also the number that cosmologists had predicted. How did they do that? By a careful study of the events expected in the Universe when it was between one second and three minutes old. This is when protons and neutrons undergo nuclear reactions to form the elements of deuterium, helium-3, helium-4 and lithium. The observed abundances of these elements agree beautifully with the predictions of the Big Bang model, showing that we have a good understanding of how the Universe is expanding just one second after the expansion began. But these predictions depend upon the assumption that there are three varieties of lightweight neutrino in Nature. If there are four then the expansion rate of the Universe is increased early on and there will be an accompanying increase in the amount of helium produced. This increase takes it above the level observed. In contrast, three neutrino types produces a helium abundance in agreement with observation.

Superstrings

Despite the appeal of GUTs, they were clearly incomplete. They failed to include gravity in the unification picture. This

Introduction

was not for want of trying; but all attempts to add gravity to GUTs met with mathematical contradictions and nonsensical results.

Then something unexpected happened. Michael Green and John Schwarz discovered a completely new type of physical theory in which these inconsistencies did not occur. In fact, its elegant consistency actually demanded that gravity be joined with the other three forces of Nature. These theories were called string theories because they were based upon the requirement that the most basic entities in Nature are not point-like particles, as had been assumed in GUTs, but were lines or 'strings' of energy. One might think of them as little elastic loops whose tension changes with temperature, so that at low temperature the tension increases and the loops contract and become more pointlike, whereas at high temperature, the tension falls and the loops become more string-like. In this way string theories promise to reproduce the successful predictions of the pointlike theories at low temperatures but make novel predictions about the behaviour of matter and the Universe at very high energies. The advent of string theories began the search for a single 'Theory of Everything' governing all the known forces of Nature. So far, string theories have proven quite easy to find but mathematically impossible to solve. Only when mathematical methods have been found which can extract their predictions will cosmologists know what they have to tell us about the first moments of the Universe.

Other speculative attempts to study the first moments of the Universe took a different approach and focussed entirely upon the unusual behaviour of gravity in the presence of quantum uncertainty. Stephen Hawking and his collaborators proposed that one should explore the possibility that the nature of time is radically altered when traced backwards towards what we have always called the 'beginning' of the Universe. Instead of this path leading to the beginning of time, we may find that time just melts away and ceases to exist. Conversely, as the Universe cools so the notion of time becomes increasingly definite so long as the starting state of the Universe was of a particular sort. If such a picture is correct then it offers the exciting prospect that our experience of a clear notion of time in the Universe today is telling us something about the nature of the initial state of the Universe.

Inflation

Studies of the very early Universe have been dominated by the idea of 'inflation', first proposed by Alan Guth in 1981. This phenomenon appears to be almost inevitable regardless of how the Universe began and offers a way of realising the dream of explaining the present structure of the Universe without recourse to special assumptions about how it began. What is inflation? Many different versions of the original theory of inflation have been proposed. At root all that is required is for the Universe to undergo a brief period of its very early history during which its expansion accelerates. This is more far-reaching than it sounds because, in the standard picture of the Big Bang, the Universe decelerates at all times, regardless of whether the expansion continues forever or will one day reverse into contraction. A period of accelerated expansion requires the existence of forms of matter that antigravitate. Whilst no such forms of matter are known in the Universe today they are predicted to exist in most theories of high-energy physics. Moreover, their response to gravity varies with

Introduction

time, so that when the Universe starts expanding they can exert gravitational attraction, so the Universe decelerates, but then they become gravitationally repulsive for a short period, during which the Universe accelerates (or 'inflates'), before decaying away into radiation, leaving the Universe to resume its decelerating expansion.

Cosmologists have studied the multitude of ways in which this sequence of events can occur and what the detailed consequences will be. The accelerated expansion enables the entire visible part of the Universe today to have expanded from a far smaller region than if the expansion had always decelerated. This enables us to understand why the observed part of the Universe is so smooth and why it expands at almost the same rate in every direction. It is the expanded image of a region that was small enough to be kept smooth by microscopic frictional processes. Another consequence of a sufficiently lengthy period of inflation is to drive the expansion of the Universe closer and closer to the critical divide separating indefinite future expansion from eventual recollapse. This offers an explanation for an otherwise most improbable coincidence about the way the Universe expands. Inflation predicts that we should find our Universe expanding tantalizingly close to the critical divide separating indefinite future expansion from collapse to a "Big Crunch", although it cannot prescribe on which side of the critical divide it lies. This means that it must contain a total density that is close to the so called 'critical' density—the largest density the Universe can have and still expand forever.

Introduction

Dark matter

If we total all the observed contributions to the density of the Universe, then we find barely one-tenth of the near-critical level predicted by inflation. This tells us that inflation predicts that a large amount of dark material exists in the Universe over and above what is required to account for the motions of stars in galaxies and galaxies in galaxy clusters. Moreover, the agreement between the observations of helium, lithium and deuterium in the Universe and the outcome of the nuclear reactions which occur during the first three minutes of the expansion tells us that the density of matter taking part in these reactions cannot be larger than a tenth of the critical density. So this mysterious dark matter that inflation theories predict must be in a form that is unlike the ordinary nuclear matter of which we are made. More specifically, it must be a form of dark matter that does not take part in nuclear reactions.

The favoured candidates for this unseen material are neutrinos or other neutrino-like particles that are required to exist if supersymmetry holds in Nature. If any of the three known types of neutrino possesses a tiny mass, little more than a billionth of the mass of a proton, then they will be produced in such numbers during the early stages of the Universe that they could provide the dark matter. Unfortunately, neutrinos that are this light interact so feebly with matter that there is no hope of ever detecting directly the sea of cosmic neutrinos all around us. More attactive has been the possibility that the dark matter is composed of much heavier neutrino-like particles, with masses close to, or exceeding, that of the proton. These weakly interacting massive particles (or WIMPs as they are acronymically dubbed) are less copiously produced in the Big Bang but that lack is more that made up for by their larger mass. Their most interesting feature is that, if they do constitute the

Introduction

dark matter in the Universe and they provide the critical density level that inflationary theories lead us to expect, then the sea of WIMPs around us should very soon be detectable by small underground detectors. As these particles rush through the Earth they will collide with the nuclei of elements like silicon in distinctive ways that leave tell-tale signatures. These can be seen so long as the bombardment of the apparatus by cosmic rays and local sources of radioactivity can be shielded out or discriminated from. During the last few years the technology required to build sensitive dark matter detectors, cool them close to absolute zero and shield them deep under the Earth's surface, has been developing rapidly. At present, preliminary experiments have exluded some classes of WIMP particle. Within a few years these experiments should tell us whether or not the Universe is awash with heavy neutrino-like particles never before detected on Earth.

Back to the beginning

Of perennial interest to students of cosmology is the issue of the beginning of the Universe—the so called initial 'singularity'—was there one and what was it like? We devoted a considerable amount of space to the subtle question of what we mean by a singularity and what one needs to know about the Universe in order to deduce its existence in the past. Much that we know flows from the singularity theorems first conceived by Roger Penrose. They dictate that if certain conditions are met then there is a singularity in our past. One of these conditions is that the expansion of the Universe always be decelerating. Remarkably, this is precisely the condition whose denial

Introduction

lies at the foundation of the inflationary universe theory. Therefore, if inflation occurs in the early universe, the singularity theorems are inapplicable and we cannot say whether or not there was a beginning to the expansion. Some inflationary universes have initial singularities; some do not.

The other area where the inflationary universe theory makes contact with observational astronomy is through its ability to produce tiny variations in the density of matter in the Universe from place to place during the first moments of its expansion. For the first time cosmologists do not have to assume the existence of some particular pattern of fluctuations built into the initial state of the Universe. When the short period of inflationary expansion comes to an end in the early universe it leaves a particular pattern of tiny variations in the density of matter which amplify as the Universe expands and eventually condense through a complex sequence of events into the galaxies we see today. But before they grow intense enough to begin the complicated process of forming real galaxies they leave a gravitational fingerprint in the sea of microwaves leftover from the Big Bang. Those fingerprints were discovered by the COBE satellite in 1992. The ongoing study of their detailed form dominates cosmology today.

Prologue

If paradise is the state of ultimate and perfect symmetry, the history of the "big bang" resembles that of "paradise lost." For the briefest instant at the origin of time when all laws of physics were on an equal footing, all nature's elementary constituents, heavy and light alike, interacted freely and democratically. The most exotic particles known, or even dreamt of, by man were liberated to participate in this unrestrained interchange. The universe was once so hot that no particle attained any permanence. All lived and died in the briefest flash of splendor. When some decayed, others appeared instantaneously to take their place; so was energy shared and shuffled about.

This idyllic era, however, was doomed to transience. Once the temperature began its inexorable fall, the symmetries were broken. Paradise was irretrievably lost; rigid patterns and diversity reigned. New particles no longer rose in the shadows of their peers. Decadence dominated the subatomic

Prologue

world, and the result is the varied universe of broken symmetry that now surrounds us.

One of the most extraordinary things about our universe is that although it often appears, at first sight, to be perfectly symmetric, closer examination invariably reveals that the symmetry is not quite exact. The universe is almost, but not quite uniform over its largest expanses; elementary particles are almost, but not quite the same as those that are their mirror images; protons are almost, but not quite stable. Could it be that things are all constructed along lines reminiscent of the ancient world where craftsmen refrained from creating patterns with perfect symmetry lest the gods be offended? No, we invariably find that the tiny breaches in the perfect pattern we might have expected to find are the cogs of a glittering mechanism at the center of things, and one of the reasons our very existence is possible.

The neutrino, a ghostly spinning elementary particle, carries with it not only an uncanny reminder of a time when symmetries were perfect but also a clue as to how they came to *be* broken. For every neutrino that now spins to the left, there was once, near the "big bang," a neutrino that spun to the right. Around us now, we see only left-handed neutrinos; their right-handed partners did not survive the early stages of the universe to emerge into the present era of broken symmetry. It is not only neutrinos that fail to be ambidextrous. Even biological development on earth is a breaking of perfect symmetry. The essential biological molecules, the helical building blocks of life, are left-handed spirals. We find only the left-handed amino acids in living things, never their mirror images. This tale of broken symmetries extends from the beginnings of time to the here and now.

Acknowledgments

We would like to thank the following individuals for reading various chapters of our manuscript and making many constructive suggestions that helped us improve our text: David Bailin, Roman Juszkiewicz, Alex Love, Stephen Siklos, and Robert C. Smith. Martin Kessler and Theresa Craig of Basic Books provided invaluable and enthusiastic editorial guidance. We are also grateful for the secretarial assistance of Jane Bamford, Julia Gilham, and Denise Haynes. Finally, our long-suffering families deserve our warmest gratitude in return for the many hours and days during which we have abandoned them because of our involvement in this book.

CREDITS

The Left Hand of Creation

1

Cosmos

A great controversy once raged among philosophers, divines, and men of science. Should Adam be depicted with or without a navel? Draw him with a navel, and you presume he had a natural mother; draw him without a navel, and he will not be a complete man. The debate gradually abated with the realization that creating a navel was a miniscule task compared to that of creating Adam. Solve the latter problem, and the other solves itself.

The earth's origin is hidden in its geological record. From the study of fossils the earth is estimated to be several billion years old. When such estimates and proofs of the earth's great antiquity emerged in the late nineteenth century, they created a crisis of belief for many who thought the earth to be only thousands of years old. The dilemma inspired a zoologist, Philip Gosse, to attempt to resolve the quandary created by the gross discrepancy between the apparent evolutionary and geological ages on the one hand, and theological prejudices on the other. In his book, *Omphalos,* he suggests that the earth, along with its entire fossil record, was created a few thousand years ago and only has the *appearance* of very great age. Not surprisingly, even the Victorians were not eager to embrace Gosse's idea of a Creator who performed such cosmic sleight of hand.

An equally extreme view of the age of the cosmos is associated with the creationist school. Inspired by Bishop Ussher, a seventeenth-century clergyman, a host of modern Christian fundamentalists provide pseudo–scientific reasoning in favor of an age of 6,000 years for the universe.

Modern cosmology has a theme that parallels the question of the earth's age. Was there a unique event, the "big bang," from which our universe began with some definite starting conditions, or is the universe but one segment of a possibly infinite series of cycles? To answer this, one approach is to seek out the oldest, most primitive stars and galaxies, and then look for the umbilical connection: What came before, if anything? No star suddenly vanishes without trace, but like the grin on Lewis Carroll's Cheshire Cat, there is always some tell-tale residue of radiation and of matter. Scrupulous astronomical detective work may some day direct us to the very first stars or to their remnants.

Age of the cosmos

It is far from apparent that the universe had an origin at a finite time in the past. Around us, stars are born and stars die. We witness the process of star birth within the murky depths of interstellar gas clouds. The death throes of a massive star climax with the outburst of a supernova. As much light is radiated within a climactic few months from a supernova as is emitted over the entire lifetime of a star. By human standards, a star's life span is long. More than a million years elapse before a star exhausts its supply of nuclear fuel and finally explodes. This may exceed the time span of the human species, but it is a mere wink on the cosmic time-scale.

Yet the material out of which stars are made, great clouds of hydrogen gas and dust, is also a limited resource in our galaxy. No matter how one does the calculation, whether by seeing how long it will take to deplete all the interstellar matter, or by inferring the ages of the longest-lived stars around us from the rate at which they are consuming their internal nuclear fuel supply, one inevitably arrives at the following conclusion: Our galaxy, the Milky Way, has only existed for a finite time, some fifteen billion years or so. It may live for many more aeons, until its stars are mostly burnt-out embers, but our galaxy undoubtedly formed in the finite past and our neighboring galaxies must have also formed at a similar epoch.

Cosmologists are resourceful individuals, however, never fearful of allowing ugly facts to destroy beautiful hypotheses. Thus it was once considered respectable to assert that the universe could be in a steady-state of infinite age. To make this compatible with the observed expansion of the universe, one has to postulate that matter is perpetually being created. New galaxies continuously develop to replace their fading predecessors. The philosophical motivation for this reasoning retains elements from antiquity. But Hellenistic philosophers had a far easier time convincing themselves that the universe was unchanging and eternal than the modern steady-state cosmologists who had to cope with a most extraordinary fact. From the unlikely realm of a dauntingly complex mathematical theory, there emerged in the 1920s one of the greatest predictions of our century: the expansion of the Universe. Within a decade, this prediction was verified. Galaxies were found to be flying apart from one another. The rate of their recession implied that the expansion began less than twenty billion years ago. This apparent beginning for the universe struck at the heart of any hopes for a universe of infinite age in a steady-state.

The greatest recession

It took the converging of two great research efforts in the 1920s stemming from theory and observation before the expanding universe hypothesis was confirmed. The characters involved seem the most unlikely protagonists. Alexander Friedman, a Soviet mathematician who spent much of his time as a meteorologist, became fascinated with the hydrodynamics of air circulation. Not content with a mere theoretical understanding of air, he proceeded to immerse himself in it with extraordinary vigor. He was an enthusiastic balloonist, and in 1925, with his pilot, attained the record altitude of 23,000 feet. Rapidly mastering Einstein's first papers on gravitation and cosmology, Friedman uncovered a major oversight in Einstein's calculations. In 1922, he found that the simplest model of the universe was in a state of either uniform expansion or contraction. (Einstein never forgave himself for the prejudice that had led him to conclude that the universe necessarily had to be static.) But Friedman died prematurely, and the expanding universe theory was promptly forgotten.

A Belgian cleric and mathematician, the Abbé Georges Lemaître, rediscovered Friedman's expanding cosmology in 1927. His work was inspired by a seminar given by Edwin Hubble on the *red shifts* of the galaxies.

By using the analogy of a speeding train, one can more easily understand the nature of red shifts. If we are standing near a railway track, the pitch of the train whistle will rise as a train approaches, but fall as the train passes and recedes. While the train is approaching, the sound waves reach our ears at a higher frequency than when they are being emitted, whereas when the train is moving away from us the sound waves are received with a lower frequency. As the train passes there will be a noticeable change in the pitch of the whistle. The change in the received frequency of waves from a moving source is called the Doppler shift and a measurement of the amount of frequency shift allows the speed of the source to be determined. This is the principle upon which the Highway Patrol's radar speed-traps are based.

Reddening of the galaxies

The stars within distant galaxies emit light that possesses a particular set of energies, or frequencies. Hubble measured these, and on comparing them with the frequencies obtained from similar materials sitting at rest in his laboratory, he found that the stellar light frequencies shifted in one direction along the frequency scale towards the red end of the visible frequency spectrum. He interpreted this shift as a Doppler effect and hence was able to calculate the recession velocities of the light sources. Hubble had discovered the expansion of the universe, and the shifting of the spectral frequencies is now known as the red shift.

The favored theoretical picture of cosmology that had confronted Lemaître was Einstein's original theory of a static universe, unchanging in time. In 1917 the Dutch astronomer Willem de Sitter offered a slightly different scheme. The curious feature of de Sitter's universe was, while it appeared to be static, the light from distant points became progressively redder. By a mathematical trick, this effect was interpreted as the result of the systematic motions of distant points. The greater the distance, the greater the red shift. Unfortunately, de Sitter's universe was empty: it was totally devoid of any matter. Einstein's world was motionless matter; de Sitter's was matterless motion, but Friedman-Lemaître's was matter in motion. Endless debates began over whether red shifts predicted by some of the theoretical universes were to be interpreted as indicators of true recession or as mere signals that light rays lost energy on their long journey through a static universe to our telescopes.

Confrontation

Meanwhile, Hubble was measuring more and more red shifts and firmly establishing the distance scale to the galaxies. Hubble (having already rejected a career as a prizefighter and found time to practice as a barrister) was a late-comer to astronomy, yet within a decade, he had changed our entire outlook on the universe. Hubble's law, the observation that the recession speed of a remote galaxy increases in proportion to its distance away, was announced in 1929 (see Figure 1.1). Even then, it took an astronomer of great vision, Arthur Eddington, to link the predictions with the proof. Eddington,

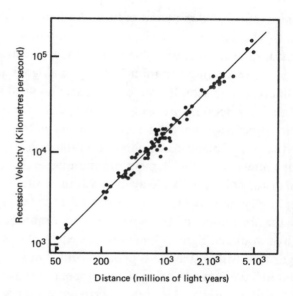

Figure 1.1 *Hubble's Law.* Observational data demonstrating the linear dependence of a galaxy's velocity of recession with distance from the observer.

a forceful and eloquent popularizer of cosmology, quickly revived Friedman's half-forgotten theory of the expanding Universe.

The expanding universe soon became a paradigm, but it was not without its critics. One of the most troublesome issues was the apparent age of the universe. The age of the solar system appeared to exceed Hubble's expansion age of the universe.

In determining the earth's age, it is important to note that the natural world possesses a number of built-in clocks. A very useful one is the process of radioactive decay. The ticks of this "clock" are the decay of individual particles. Some take so long to decay that they can be used to mark time over cosmological epochs. One pound of uranium-238 is reduced by half into various decay products after one billion years. If we measure the amount of stable debris, then we can infer the age of the remaining uranium-238. By analogous techniques, different rock samples can be compared to yield their ages. With considerable precision independent studies of such radioactive fossils have determined the age of the oldest rocks in the solar system to be about 4.6 billion years. The precise age inferred for the element uranium depends on the chemical history of the Milky Way in its youth. If the synthesis of heavy elements like uranium occurred at a steady rate prior to the formation of the solar system, uranium's estimated age would be eight billion years. In effect, this would be the age of the Milky Way in its present form. The uranium is believed to have formed during supernova explosions when the collapse of a dense stellar core liberated vast numbers of high energy neutrons that irradiated the outer layers of the star. We do not know whether supernovae occurred at a more prolific rate in the early stages of the Galaxy's life. If they did, a somewhat greater age for the uranium would be inferred.

Hubble's initial determination of the age of the universe from the galactic recession was only two billion years. We know now that his estimates of distances to the galaxies were grossly and systematically in error. Another two decades would pass before improved techniques for estimating the distances would resolve this age discrepancy. In the meantime, the "big bang" adherents put their faith in the simplicity and elegance of the theory. We now infer from the ratio of distance to velocity for receeding galaxies that the explosion was initiated about fifteen billion years ago—if we assume that the expansion velocities have been constant during this time (see Figure 1.2). Some deceleration could have

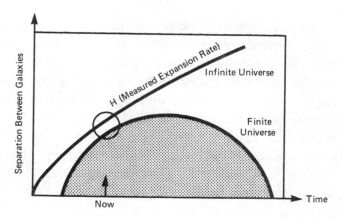

Figure 1.2 *Expansion of Cosmic Distances with Time.* In an infinite universe, the separation will continue to grow indefinitely; but, in a finite universe, the expansion will one day be reversed into contraction. The age of the universe can be determined in both cases by the rate of expansion that is measured today.

occurred and even acceleration is possible. But the remarkable agreement between the general magnitude of the cosmological time-scale and the age of the solar system (and also the ages of the oldest stars in our galaxy) shows such effects to have played only a very minor role in the evolution of the universe.

Alternatives

Avoiding the dramatic conclusion that the universe is expanding has been the preoccupation, and even the obsession, of numerous cosmologists. The "tired light" hypothesis was one early idea. Perhaps photons, it was thought, simply lost energy on their long journey to us through intergalactic space. Some simple observations, however, eliminated this idea. One would expect that the degree of photon fatigue should depend on the photon frequency. But we can precisely measure identical red shifts at frequencies as disparate as radio wave frequencies at 10^9 Hertz and frequencies of visible light at 10^{15} Hertz. Tiring light would result in increasing amounts of blurring, either of images or of spectral lines, with red shift. This phenomenon is not observed, implying that photons, particles of light, simply do not lose energy in transit to our telescopes. Another possibility is that the red shift of light has a gravitational origin. Indeed, photons do lose energy in overcoming the gravitational force of a dense object as they escape from it. The effect has been measured on earth, where it amounts to a red shift of 10^{-9}.

In the Milky Way galaxy the gravitational component is larger, causing the light from a typical star to be red-shifted by 10^{-6}. The actual red shift of the most remote galaxies, however (approximately one or so), is so much greater than any possible gravitational red-shifts that no model of a galaxy could yield a red-shift value that was due to gravity alone.

Steady-state

The "big bang" theory holds that creation occurred at a single point in time. To many cosmologists, it once seemed equally plausible that creation occurred everywhere in space throughout all of time, at just the rate required to overcome the dilution by expansion. The newly created matter condensed into galaxies, and the universe never changed its appearance. There was no beginning.

This steady-state universe originated one night in 1946 when Hermann Bondi, Thomas Gold, and Fred Hoyle saw a movie of a ghost story that was ingeniously designed to have a circular plot with the end identical to the beginning. "What if the universe is constructed like that?" queried Gold over brandy later that night in Bondi's rooms overlooking Trinity Great Court in Cambridge. From that curious beginning the steady-state theory of the universe was born. It was published in 1948. Although Hoyle's original paper was rejected by the Physical Society of London "on account of the acute shortage of paper," the eventual popularization and publicity surrounding the theory forced it into the public imagination. This only encouraged debate.

The issue of "big bang" versus steady-state dominated the cosmology of the 1950s. The steady-state proponents eliminated the need for an origin of the universe in time by a daring artifice that raised the hackles of people in certain quarters. It may have been, in historian Stanley Jaki's words, "the most daring trick ever given scientific veneer."

The steady-state concept of *creation ex nihilo* blatantly stole an official part of Christian dogma to serve an almost non-Christian goal, but the notion of continuous creation throughout all of space served aptly to counterpoint the opposing theory of a unique creation in the remote past. The battle for a steady-state universe of infinite age and continuous creation was waged in the 1950s and lost in the 1960s. It should serve as a reminder of the potential pitfalls that await any theoretical model of the universe.

The Left Hand of Creation

The night sky is dark

Actually, a very simple observation, which requires no
detector more powerful than the human eye, demonstrates
the finite age of the universe. Practically every form of matter
emits radiation. If matter is cool, the radiation is at infrared
or even microwave frequencies. If it is hot, the radiation is
visible at ultraviolet or x-ray wavelengths. All of this radiation is emitted into space and accumulates in the vast cosmic
reservoir of the universe. Through telescopes the radiation
appears as a faint diffuse background glow. The night sky
does not, even on the most remote mountaintop, appear
totally black between the twinkling stars. The unaided
human eye may not discern the cosmic background light, but
photographic plates reveal its presence. One finds cosmic
background radiation over the entire span of the electromagnetic frequency band. Generally, it appears to be the cumulative light from vast populations of stars, poured out over
billions of years. But, it is still exceedingly faint, amounting
to no more than 1 percent of the brightness of the Milky
Way.

The obvious fact that the sky is dark at night is a good deal
more subtle than it seems. If the universe were static and
infinite, with galaxies sprinkled throughout space, then,
wherever one looked, the line of sight would end on a star.
It would be like looking into a forest of trees. Your line of
sight is always a continuous wall of trees. The entire night
sky would be as bright as the surface of a star. This paradoxi-
cal state of affairs, called "Olbers' paradox," after a nine-
teenth-century scientist Heinrich Olbers, is resolved if the
universe is finite. As we look farther out into space, we see
events as they were further back in time because of the length

of time it takes light to reach us. An age for the universe of ten billion years limits our horizon to ten billion light years. This resolves Olbers' paradox because only galaxies within this distance are able to contribute their light to the brightness of the night sky. But definitive arguments are rare in astronomy. The red shift degrades the light from distant stars, and this light will therefore be shifted from visible wavelengths towards the infrared spectral region. By using this process steady-state cosmology could also be reconciled with Olbers' paradox.

A cosmic signal

The final demise of steady-state cosmology and of other serious rivals to the "big bang" theory was due to the discovery in 1965 of the cosmic microwave background by two radio astronomers. Arno Penzias and Robert Wilson, who were later awarded the Nobel Prize for their achievement, fortuitously stumbled upon the cosmic microwave radiation while attempting to calibrate one of the first satellite communications telescopes with which they hoped to measure radio emission from the Milky Way. They found a residual signal which had the same intensity wherever they looked in the sky. This extreme degree of uniformity from one direction to another, or isotropy, was the clue that the radio waves might originate in the distant depths of the universe. In fact, back in 1948 the "big bang" cosmological hypothesis led George Gamow and his colleagues, Ralph Alpher and Robert Herman, to predict the existence of just such a low temperature radiation field pervading the universe. The predicted radiation had the character of black body radiation, heat radiation emitted by an idealized, perfectly efficient heat source (see Figure 1.3). With this theoretical backing, cosmologists soon identified the background radiation found by Penzias and Wilson as the vestigial radiation of the "big bang," although it was another decade before the black body nature of the spectrum was finally established.

The microwave frequency band straddles wavelengths between a millimeter and a few centimeters. Neither stars nor normal galaxies emit a significant amount of microwave radiation though. Indeed, to an observer with microwave vision,

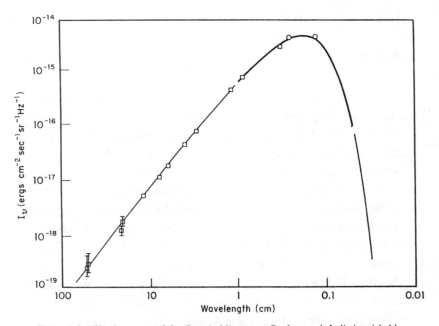

Figure 1.3 *The Spectrum of the Cosmic Microwave Background.* Indistinguishable from the spectrum of a perfect blackbody radiator, this graph shows the COBE satellite measurements (bold face) at high frequencies (or short wavelengths), where no distortion is seen to a precision of better than 0.03 percent, and the low frequency ground-based measurements where the errors are larger because of contaminating backgrounds from the atmosphere and the Milky Way.

the universe would appear virtually uniform and structureless. The Milky Way would be dim and appear primarily as a bright haze of microwave light. There are no known sources bright enough at microwave wavelengths to account for this cosmic background. Furthermore, the background is so uniform in intensity from place to place that it would require more sources than there are galaxies in the entire universe to keep their random intensity fluctuations down at the observed level. The only conceivable origin must lie in the remote depths of the universe, and radiation must be as fundamental a constituent of the universe as matter itself.

Black body radiation

The intensity of microwave radiation background is found to peak in a manner characteristic of black body radiation at a wavelength of about two millimeters. To the physicist, a black body is a perfect radiator. The radiation comes from matter that is in complete thermodynamic equilibrium. One can create this state by constructing a cavity and allowing no radiation to enter or leave the cavity. Eventually, the matter in the cavity attains equilibrium with the cavity wall. The radiation field in the cavity is then that of a black body at the temperature of the cavity wall. The colder a black body, the longer its wavelength of peak intensity. The cosmic microwave background radiation has a temperature of just three degrees above absolute zero. Remarkably, seventeen years earlier, Ralph Alpher and Robert Herman had predicted its value would be close to five degrees above zero, almost exactly what was found. However they did not realize it should be experimentally detectable. The credit for this crucial step goes to two independent groups. In 1964, two Russian scientists, Igor Novikov and Andre Doroshkevich, came up with this proposal. Then in 1965, Robert Dicke and his coworkers at Princeton rediscovered Gamow's hot big bang theory, and even set about measuring the left-over radiation themselves then, only to be beaten in the race by Penzias and Wilson's serendipitous discovery.

A black body radiation field is described uniquely and completely by its temperature. At three degrees Kelvin, that is -270 degrees centigrade, there are about $5 \cdot 10^8$ photons of average energy $3 \cdot 10^{-4}$ electron volts in every cubic meter. In the same volume there are only, on average, about 4 protons, each of which has mass equivalent to an energy of a billion electron volts. Allowing for dark matter we have not yet seen may increase this by a factor 10 or so, if the dark matter consists of protons—rather than say, neutrinos (elementary particles that have almost no mass and travel at the speed of light) or other unknown particles. Consequently, the energy density of radiation is less than 1 percent of that of the luminous matter. But this statement applies only to the present state of the universe. As we travel farther out into space and consequently back in time, the presence of radiation becomes increasingly more significant. This is because the temperature of the black body radiation falls as the universe expands. Consequently, the mean energy of a photon has been steadily decreasing. If we think in terms of the wavelength of the radiation, the reason for this energy decay becomes apparent. The wavelength of black body radiation is, on the average, about equal to the separation between photons; they are as closely packed as possible. As the universe expands, the wavelength increases, and the increase is directly proportional to the increase in scale of the universe due to its overall expansion. Accordingly, we can say that the frequency of the radiation, and, therefore, the energy of an individual photon, decreases proportionately as the scale of the universe increases with time.

The implication of this reasoning is rather astonishing. It means that when the scale of the universe was less than one thousandth of its present value, the dominant constituent of the universe was the black body radiation. It is this primeval state in which radiation predominates that provides the setting for the unfolding saga of the early evolution of the universe.

Disorder

Energy density is not the only useful property of the black body radiation for the cosmologist. We shall see that entropy, a measure of the disorder of the radiation field, provides another useful concept. The entropy for pure black body radiation is approximately equal to the number of photons within the cavity containing the radiation. At high temperatures, the entropy is much higher than at low temperatures for a fixed volume. As natural machines, stars produce large amounts of entropy. Hot photons stream out from their core and are degraded into colder photons at the stellar surface. In cosmology, since the only natural volume is that of the entire universe—which could well be infinite—the total entropy is not a very meaningful concept. It is more useful to speak of the amount of photon entropy per individual proton. We call this the specific entropy. It is roughly the ratio of photons to all heavy subatomic particles such as protons and neutrons, which we collectively call baryons, within an arbitrary volume. The specific entropy at any point provides a measure of the local entropy. Since the number of black body photons is proportional to the volume, as is the number of baryons, the specific entropy is a constant—a pure number—as the universe expands, provided that photons and baryons are not being created or destroyed. We shall see that this condition is well satisfied over a considerable range in time. The present value of the photon to baryon ratio is about 10^{10} if we count only the luminous matter content of the universe when calculating the number of baryons.

This ratio is huge compared with the specific entropies of familiar systems of radiation and matter and even those encountered in common astrophysical objects like stars or nebulae. A candle flame and a fluorescent strip of light have specific entropies of about 2 and 1000 respectively; the amount of entropy generated by the sun during its entire lifetime is about 10^6 photons per baryon. A supernova explosion generates somewhat more entropy, perhaps as much as 10^7 photons per baryon, still well short of the cosmic value. This immediately tells us that the cosmic background radiation could not have been produced by astrophysical sources that resemble the objects we now see in our galaxy. Some far more exotic source must be sought. The "big bang" cosmology provides such a source. A nonprimordial origin for the cosmic black body radiation cannot be completely discounted. The alternative models to the "big bang" however, invariably require some coincidences between otherwise quite arbitrary parameters, a state of affairs abhorrent to most physicists. The simplest explanation, the minimal hypothesis, is provided by the "big bang" theory. It is not necessarily the whole truth, but its straightforward explanation of numerous facts argues that it is the best approximation to the truth that we currently possess.

Uniformity

Now imagine comparing the density of radiation to that of matter as we move around in the universe. We would discover that while the radiation remains remarkably uniform, the matter distribution, at first sight, appears to be extremely lumpy and inhomogeneous. In fact, uniformity or homogeneity are relative concepts. Examine a seemingly smooth surface with a microscope and its smoothness vanishes. On a dark night, the sky is filled with a million or more visible stars. They seem to be more or less sprinkled at random over the sky. This, too, is an illusion that evaporates upon closer scrutiny. Look through a telescope, and you find that practically all the bright stars are near neighbors to the sun and belong to our Milky Way galaxy. As the telescope power is increased, the rest of our galaxy comes into view as a distinct island of stars and nebulosity. One has to penetrate two million light years into space to encounter the next bright galaxy, Andromeda.

As the telescope power increases, we see deeper and deeper into space. Thousands of galaxies come into view; some are spirals like Andromeda and our Milky Way, others are ellipsoids resembling oversized globular star clusters. We are now peering to a depth of one hundred million light years. The galaxies are mostly concentrated in the vicinity of Virgo. We are seeing the Virgo cluster and supercluster of galaxies.

Our own galaxy, within its own Local Group of companion galaxies along with Andromeda, is an outlying member of the Virgo supercluster.

Finally, with the aid of the largest terrestrial telescopes, we can look one, three, even ten billion light years into space. Now we see galaxies by the million. Indeed, as many as a billion are now detectable from earth. There are many thousands of groupings and clusterings, even occasional dark voids empty of bright galaxies. There actually appear to be both bridges and sheets of galaxies. But no longer can we discern any region where the concentration of galaxies is greatest. The galaxies are evenly distributed throughout the universe. Faint galaxies are scattered everywhere like specks of dust, outnumbering by far the larger bright galaxies. Indeed, there seems to be almost no end to what one recognizes as a galaxy, ranging from gigantic ellipticals brighter than one trillion suns to sparse dwarfs with a mere hundred thousand suns that are barely visible above the glow of the night sky. The population of the universe is diverse indeed.

Isotropy

In every direction there are galaxies. The universe exhibits no bias for one direction over another. There is no discernible edge or center. The galaxies are swimming in a uniform sea of cooled radiation—the cosmic black body radiation. The uniformity of its temperature provides the most overwhelming evidence for the homogeneity of the universe. No variation in its intensity larger than one part in a thousand is seen from one direction in the sky to another. This uniformity alone is compelling evidence that it originated far away, in the remote depths of the universe. As we have seen, the weakness of any microwave emission from nearby galaxies demands an origin shrouded by mists of pregalactic time. Only the "big bang" itself, or some exotic process far removed in space and time, can account for the cosmic black body radiation.

The cosmic black body radiation is anchored to the universe. It provides a cosmic frame, relative to which we can measure the earth's motion, as it orbits the sun and the sun orbits the Milky Way (see Table 1.1). But this is not the end of the hierarchy; even the Milky Way is moving in what is called our Local Group around a point roughly midway between us and Andromeda. Even the Local Group is on the move, responding to the gravitational pull of the great Virgo supercluster of galaxies. This attraction decelerates the relative separation of Virgo and ourselves that is created by the Hubble expansion of the universe. Consequently, our motion away from Virgo is reduced by about 25 percent relative to what it would be were our local environment truly uniform.

Table 1.1 *Our Motion*

Motion of earth relative to cosmic black body radiation	410 kms^{-1}
This includes the following known components:	
Earth relative to center of gravity of earth-moon system	0.013 kms^{-1}
Earth-moon system relative to sun	29.8 kms^{-1}
Sun relative to local standard of rest defined by nearby stars	20 kms^{-1}
Local standard of rest rotation around the galaxy	250 kms^{-1}
Center of the galaxy relative to the Local Group of galaxies	100 kms^{-1}
Local Group relative to Virgo supercluster	300($\pm$100) kms^{-1}
Virgo Supercluster realtive to Great Attractor	400($\pm$100) km/s

Our net motion through the black body radiation has been measured. Directly ahead of us, in the direction we are moving, the radiation rushes toward us and is measured to be fractionally more intense than from other directions. Similarly, in the direction away from which we are moving, the radiation is found to have a slightly diminished intensity. The effect, amounting to about one part in a thousand, is actually due to the Doppler shift. The wavefronts that separate successive wavelengths of the radiation are slightly compressed in the forward direction as a consequence of our continuing motion during the interval between the arrival of each wavefront. This shift to shorter wavelengths, or blue shifts, results in a slight increase in intensity of the black body radiation, since a decrease in the wavelength corresponds to an increase in photon frequency and energy.

Whether the gravitational pull of the Virgo supercluster suffices to account for the entire magnitude of our cosmic motion cannot yet be definitively ascertained. Certainly, it plays the dominant role, although more distant irregularities in the galaxy distribution may also contribute to the acceleration of the Local Group of galaxies. [1]

A fossilized fireball

The black body radiation is a fossil of the primeval fireball from which the universe began. For ten billion years this radiation has streamed freely throughout the cosmos. Today, it provides us with a unique record of the portions of space and time it has traversed. When examining this radiation, we are peering into the remote past when the universe was opaque. Conditions in the early universe resembled those at the center of a star. Only after a million years of expansion and cooling did matter rarefy sufficiently for space to become transparent to the black body radiation. By that time, the primordial heat had abated sufficiently for the electrons and nuclei to form atoms. Prior to this, the state of matter was plasma, containing ions and electrons. Electromagnetic radiation consists of oscillations in the electric and magnetic fields that interact with the electric fields of any free electrons. This interaction results in a scattering or change in the direction of the wave. In an atomic or molecular gas, where there are very few free ionized particles, electron scattering of the radiation is largely suppressed. Some three hundred thousand years after the "big bang" marked the epoch of the last scattering, the temperature had fallen to about three thousand degrees Kelvin and practically all the electrons had combined with protons into atoms of hydrogen. Somehow, apart from this primeval amorphous haze of atoms and photons, galaxies and stars managed to condense.

The universe is a surprisingly complex place. There are more stars whirling around our modest Milky Way galaxy than there are people who have ever lived on earth. More stars than one could even count in a lifetime, or in a thousand lifetimes. Yet, there are as many galaxies again scattered throughout observable space. The sheer number of stars in the universe is so large that it defies our imagination. Only in one other situation do we confront such vast numbers. A sugar lump has as many atoms as there are stars in the observable universe. Microcosmos here confronts macrocosmos, seemingly by chance. But elsewhere there are deep and fundamental links between elementary particles and the universe. Long ago, the seeds of cosmic evolution were sown by the interactions between atomic nuclei and their constituent parts and the force of gravity. Whatever happened then, within a matter of seconds during a dense and hot phase that characterized the very early universe, has influenced all later phenomena. Every galaxy, every star, every planet, owes its existence to primordial events whose nature is partially obscured by the passage of time. A modern perspective on cosmic evolution leaves little to chance, until the first habitable planets emerge some billions of years later.

Origin of structure

Our goal is to unveil the shrouded secrets of the earliest instants of the universe and to elucidate the cosmic connection between the beginnings of the universe and what we observe around us today. The link is not always direct, and our path will often seem to descend tortuously into murky realms of partially understood physics before attaining its destination. Inevitably, the origin of the universe lies at the frontiers of modern physics, but this need not deter us from our quest.

Space and time are inextricably linked in cosmology. Distances probed by our telescopes are so vast that the time for distant light from galaxies to reach us requires practically all of the time since the "big bang." Cosmology attempts to explain the origin of what we see in terms of things we cannot see but which we regard as plausible, simple, and natural. Our vision of outer space is limited by a natural horizon, prescribed by the distance light has travelled since the "big bang." It defines what we would call the "observable universe.'

Horizons

This horizon encompasses a billion or more galaxies. As the universe expands so the horizon size increases. Once, before any galaxies had formed, there was very little matter within the horizon of any hypothetical observer. So little time would have elapsed since the "big bang" that the horizon would have contained far less material than a typical galaxy does today. At the earliest moment of the universe, the horizon only encompassed a millionth of a gram of material. This ultimate instant is determined by the breakdown of our gravitational description of space and time. (A quantum description of gravity is needed, and this is lacking at present.) All the material and energy we can now see was present then but spread over many, many regions of horizon size. Matter in any one region had no causal relationship with that in any other part of the universe more than a horizon scale away. The beginnings of the galactic structures we see today were laid down at this threshold of cosmology. To look beyond this moment, to delve into the mystery of the initial "big bang" singularity, unarguably poses physics with its greatest challenge.

Cosmic evolution

Our story commences where stories usually do, at the beginning. First, we describe what modern cosmologists have come to think the "big bang" singularity was and the uncertainties surrounding it. Moving forwards in time we explore the unusual processes expected in its vicinity and the limits placed upon the applicability of our theories of gravity and matter at high density. In chapter 3, we move into the realm of classical gravitation, where our knowledge of particle physics, although incomplete, does not prevent us from learning how the diverse composition of the matter around us was created. It took not a week, but less than a second for practically all of the matter in the universe—precisely 98 percent of it—to form. The origin of the remaining 2 percent contains the story of man. Billions of years were required before this evolution was culminated in stagnant pools pierced by lightning flashes from the nauseating, methane-rich atmosphere of a forming planet. In chapter 4, the story of these billions of years is described. The initiation of the seeds of the first embryonic galaxies to develop occurred very early, certainly within seconds after the "big bang." Only much later did the condensation of recognizable galaxies, clusters, and superclusters of galaxies, and ultimately the stars and planets take place.

All that we observe around us is linked inextricably to conditions in the exceedingly remote past. That is our theme. But how uniquely can we pin down what the beginning must have been like? Was it chaotic or smooth, hot or cold? Chapter 5 is devoted to a careful reexamination of some of the issues that have plagued cosmologists over the years. Some models of the universe have unpredictable consequences; these clearly do not fall into the grand pattern that we have sketched. Others may not be congenial to the evolution of man, again, a disaster of philosophical, if not of cosmological, impact. Of course, many conundrums remain, and we present a potpourri of issues and answers in our final chapter.

No holds are barred; we range from the origin of time to its end, from deity to time-travel.

As we approach and run into the frontiers of knowledge, our journey may enlighten some and dismay others, but it will surely reveal much that we know, and do not know, about the origin to which modern astronomy is pointing.

2

Origins

IT has been said that the mark of good philosophy is to begin with an observation so mundane that it is regarded as trivial and from it deduce a conclusion so extraordinary that no one will believe it.

For centuries thinkers regarded the universe as a vast and unchanging background stage on which the regular motions of the stars and planets were observed. But Edwin Hubble's measurements of the reddening in the light from distant galaxies was confirmation of a very different universe. Hubble substantiated Friedman's radical prediction that the entire cosmos is in a state of dynamic expansion. The distant galaxies are receding away from each other at speeds approaching that of light. It is from these simple facts that we are led to an extraordinary, and perhaps unbelievable, conclusion. If we follow this expanding universe back in time to its origin, it appears that every region of the universe must have been squeezed into a single point of infinite density at one moment in time. This instant was christened the *singularity*, because the simple cosmological models cannot describe what lies before it. Judging by the current expansion speed of the universe and its slight deceleration, the cataclysmic beginning occurred less than sixteen billion years ago. For perspective, the oldest fossilized bacteria found on earth are only about three billion years old.

Inevitability of a beginning?

In the early 1930s, when the first modern cosmologists realized this remarkable consequence of an expanding universe—a singularity of infinite density in the finite past—they were understandably loath to believe it. A fascinating debate arose about the reality and nature of this singularity. The debate culminated in the mid-1960s with some real understanding of what the beginning of the universe was according to the "big bang" theory. To appreciate the answer to the question, What was the "big bang" singularity?, we must follow some of the early, unsuccessful attempts to explain or explain away the "big bang" singularity.

The singularity was first explained as the result of some pathology in the models of the expanding universe, rather than as some natural phenomenon. After all, even Newton's traditional theory describing gravitational forces allowed one to predict a point of infinite density, in principle. If a perfectly coordinated collection of masses was fired toward a single point, they would all arrive there simultaneously. While Newton's theory predicts an infinite density at that point, nothing of the sort ever happens in reality. Why?

Because the model contains some idealizations and simplifications that are not reflected in nature. Any collection of converging masses in the actual world cannot behave exactly like the perfectly coordinated and symmetrical example. Other forces of nature intervene to help avoid such singular and pathological behavior.

Some cosmologists believed that as soon as a more realistic model of the expanding universe was studied—one in which the expansion did not proceed at precisely the same rate in every direction—the awkward prediction of a singular beginning would vanish. Alas, and to their great surprise, this hope was dashed. Acting on a suggestion by Einstein in 1933, Georges Lemaître tested out this reasonable assumption on a model universe that expanded at different rates in different directions. Sadly, the problem of the singular past was only compounded. Instead of defocussing the convergence of material as it was followed back into the past, the asymmetries predicted not only that the moment of infinite density remained, but that it occurred even more recently than in Friedman's model. All manner of distorted, lumpy, and even rotating cosmologies were subsequently scrutinized; all possessed the same defect. They all predicted a beginning to the expanding universe at one point of infinite density.

Pressure

A few years later, Einstein raised another objection. Think back to our perfectly coordinated collection of masses again. Even if they had been able to meet at a single point, we know from experience that they would not become arbitrarily close or infinitely dense. The masses would simply rebound off each other like colliding billiard balls. In other words, some counter-pressure would prevent a singularity. Einstein pointed out that the naive cosmological models that had so far been created from his theory of gravity all assumed that matter exerted no pressure at all. The cosmic material was imagined to behave rather like a cloud of smoke particles than a cluster of hard balls.

We know that the assumption of negligible pressures provides a very good description of the universe today. Whatever gas there is in intergalactic space is simply not hot enough to exert any pressures of cosmic significance. While it is possible for pressures to be important locally, in galaxies and even in clusters of galaxies, they simply cannot compete with gravity over the dimensions on which the galaxy recession motions approach light speed. Only as we extrapolate backwards in time, into the moments when the expanding universe was of higher density than it is today, do we find that collisions between particles and radiation become more and more frequent. Consequently, the temperature rises, and the pressure begins to mount. Eventually, pressure forces become overwhelmingly important. Perhaps the inclusion of these realistic pressures could exorcise the "big bang" singularity? Could the universe just bounce back into a state of expansion after it had contracted? After all, pressure from the increased rate of molecular collisions within a squashed balloon causes it to rebound and regain its shape.

Unfortunately, this line of thought was also destined to fail. The addition of pressures to the expanding universe model was fruitless. Perversely, instead of removing a state of infinite density in the past, pressure merely reinforced it. Einstein's famous formula $E = mc^2$ reveals the reason. It shows that all forms of energy E (pressure is nothing but a form of energy) are equivalent to mass m. The connection is provided by c^2, the velocity of light squared. So, like all masses, any pressure must both create and respond to gravitational forces. Now we can see why pressure cannot stop the "big bang" singularity. As we go backwards in time and approach the singularity, the pressure does indeed rise enormously, but pressure gravitates too and it creates at the same time an extra gravitational squeeze that more than compensates for the increase in resisting pressure. Trying to avoid the singularity by generating pressures is self-defeating, like pulling yourself up by your own bootlaces. The gravitational pull created by the increasing pressures of the "big bang" serves only to add strength to the arguments for a "big bang" singularity.

The last objection raised against the ubiquitous singularity, which haunts the expanding universe models, was the most subtle. Its beguiling credibility led many to misjudge the nature of the "big bang" and to conclude that no one who retraced the history of the cosmos would ever encounter a real singularity in nature. Instead, it was a mirage, a mere mathematical illusion, similar to an illusion that we know of in another, more familiar, setting.

Coordinates

On terrestrial globes, geographers map a network of coordinate lines of longitude and latitude that enable us to label points on the earth's surface unambiguously. As we move away from the equator toward the North and South poles, the lines of longitude begin to converge and the meridians eventually intersect on the globe at the poles. At these two points, the system of latitude and longitude used to map points on the earth's surface develops "singularities." But if we pay a visit to the North or South Pole, we can readily confirm that no real physical singularity has arisen to rupture the earth's surface. We have created only an artificial singularity in a system of map coordinates. We could, if we chose, pick a different grid of map coordinates to cover the surface of the globe which are completely regular and nondegenerate at the poles. Incidentally, no matter what alternative grid we do choose for our map, it would always possess a coordinate singularity, where the lines degenerate and overlap, somewhere. Mathematicians call this predicament the "Hairy Ball theorem," so named because it explains why, when your hair is combed flat over your head, there is always at least one singular point or "parting" where the hair cannot be combed smoothly and from which hair appears to radiate. Combing your hair in a different style may alter the position of this cranial "singularity," just as changing map coordinates could alter where the meridians intersect, but it can never remove the intersection completely. How do we know that our persistent "big bang" singularity is no more malicious than one of these innocent coordinate irregularities, the artifact solely of a defective way of *describing* and mapping the expanding universe?

Several investigators in the early 1960s thought this was all there was to the "big bang" singularity and suggested that upon encountering a moment when the universe was of infinite density, we should simply look for a new set of descriptive coordinates that make it go away. This works for a while, but, just as with geographical coordinates, the singularity moves somewhere else. When the new set of cosmological coordinates is found to become singular, they are replaced, only to become singular again, and so on *ad infinitum.* In order to decide whether our cosmological singularity is real, we have to ask what does happen *ad infinitum:* if a real physical catastrophe is taking place, we ought to find it spoiling the description in every set of coordinates as we approach its location. How do we recognize the true nature of the situation?

The edge of the universe

To resolve these awkward dilemmas, cosmologists realized that they had to define carefully what they meant by a singularity. They needed a definition that avoided all the complications of pressures, asymmetries, and coordinates and what got to the root of the pathology in the models of the beginning of the universe. To achieve this, the traditional notion of the singularity as a place and moment of infinite density, temperature, and indeed everything else, had to be abandoned. To see why, imagine that a cosmologist is presented with a complete model universe, obtained as usual by solving Einstein's array of equations. This provides a map of a universe with all its spatial and temporal characteristics. Our cosmologist scrutinizes it under his cosmic magnifying glass for places where the density, or anything else, became infinite. When he finds these spots, he cuts them out, creating a slightly perforated space-time map with a label attached— "nonsingular universe." We would desperately, but belatedly, protest that we were being cheated and that the perforated universe model was really singular, or at the very least, *almost* singular near the excisions. However, our per-

36

forated universe is a bona fide solution of Einstein's equations. To resolve this sort of dilemma, we must decide exactly what we mean by a singularity. A more careful definition is required before we can be sure that any nonsingular model universe we find has not just had its real singular points cut out by the method we used to find it.

A singularity will be said to occur when a path of any light ray, through space and time comes to an abrupt end and cannot be extended any farther. Indeed, what could be more "singular" for a traveller through space and time than the Alice-in-Wonderland experience awaiting him on an incomplete path to the edge of the universe. At the end of this path, the traveller disappears from the universe. He runs out of space and time into some sort of limbo. Intriguingly, this definition is chosen so that if our paths through a space and time model do encounter regions where someone has pruned out all those points where the density goes infinite, we can immediately tell. The path would come to an end once it hit the boundary of a hole in the map just as surely as if it hit a real point of infinite density. In both cases, the unextendable history would signal a singularity to the cosmologist. An intrepid space-time traveller who backtracked along one of these incomplete cosmic cul-de-sacs would hit the singular boundary and just disappear from the world and our description of it. The boundary itself is not part of the universe. Conversely, particles could appear spontaneously at the singular boundary.

This picture of the edge of the universe is a very useful one. It neatly sidesteps all the difficulties about the shape and content of particular universe models and is completely general. It is also something potentially very different from our image of the "big bang" as some sort of giant cosmic explosion from a state of infinite density and temperature—a notion too vague to be useful. Our new picture is more akin to the traditional metaphysical picture of creation out of nothing, for it predicts a definite beginning to events in time, indeed a definite beginning to time itself.

The reason this new definition was adopted was not just because it is unambiguous. It is possible to deduce whether or not our universe possesses one of these singular edges in the past, a state, if you wish, before which there was nothing —neither matter nor motion, nor space, nor time.

Is our universe singular?

In the mid-1960s, Roger Penrose, an English mathematician, showed how this remarkable conclusion could be established. Subsequently, a number of similar mathematical arguments were framed and are now known collectively as the "singularity theorems." The most powerful of these was constructed by Penrose and Stephen Hawking and employs no assumptions that cannot be tested by observation. This theorem proves that Einstein's theory of gravitation guarantees the existence of a singular boundary to space and time in the past if certain assumptions are made. First, gravity is always attractive and exerts its tidal forces on everything. Second, it is impossible to time-travel back into our own past. Third, there is enough material in the universe to create a *trapped region* from which even light cannot escape. These are not unreasonable assumptions. Certainly, we always observe that gravity is an attractive and universal force. Most people would regard time-travel and the accompanying possibility of overthrowing the rule of cause and effect as something even worse than the "big bang" singularity. We could meet, and even kill, our own parents in their infancy, thereby producing all manner of logical conundrums. The reason we need to include this explicit veto on time-travel is because Einstein's gravity theory allows, although does not demand, that it is possible. But it is the last proviso about a trapped region and its physical meaning that is the crucial ingredient of the theorem.

If we ignore gravity for a moment, we know light rays travel at a constant speed and so are represented by straight lines in a *space-time diagram.* This is simply a chart showing distance travelled from the point of emission with time. The slope of two lines (one each for motion starting in opposite directions) just gives the magnitude of the speed of light. If we now include gravity, which acts on everything, even light itself, the light rays will bend and slightly focus toward each other by their mutual gravitational attraction and that of the intervening material. Obviously, this effect is incredibly weak unless there is a material object present in the vicinity, whose gravitational attraction causes the light rays to deviate significantly from straight lines (see Figure 2.1). Gravity acts like a weak converging lens and the consequent light-bending has been observed by astronomers. For example, during eclipses of the sun, stars whose images should be invisible because they are blocked from our view by the disc of the sun are visible. What is happening is that light rays from a distant star pass close by the sun and are bent by the pull of its gravity, enabling us to effectively see around corners (see Figure 2.2). It does not take very much imagination to see that if enough matter accumulates in a particular region, its gravitational pull might be strong enough to stop incoming light from ever escaping again. In this case, the gravitational "lensing" is so effective that it leads to strong focusing. We

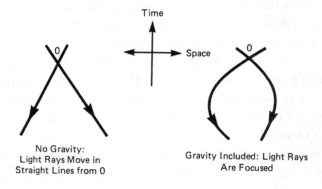

No Gravity:
Light Rays Move in
Straight Lines from 0

Gravity Included: Light Rays
Are Focused

Figure 2.1   *The Motion of Light Rays.*

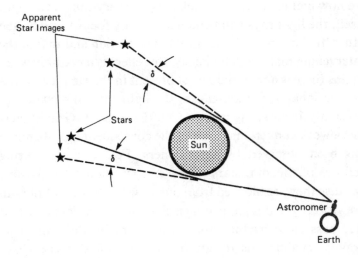

Figure 2.2   *Light-bending.* Effect of light-deflection by the sun's gravity on the apparent position of the stars. The deflection, δ, is about 0.0004 degrees.

now say there is a "trapped surface." Everything, light or matter, within the trapped surface is doomed to converge under the influence of its own gravity.

Stephen Hawking and George Ellis first realized that Penzias and Wilson's discovery of the microwave background radiation showed that the whole universe possesses a trapped surface. The total amount of gravitational attraction exerted by the microwave background itself (together with any intergalactic matter that scatters it en route) is always sufficient to create a trapped region encompassing our entire observable universe. As we trace the microwave photons back through their history, they trace out paths through space and time that are focused because of their gravitational influence on each other, as in Figure 2.2. A practical consequence of this is that the images of distant radio galaxies with similar real size should not appear smaller and smaller as they are observed farther and farther away (and so at earlier and earlier times); instead, there should exist a minimum *apparent* size.

We can confirm that the assumptions of the singularity theorems hold true in the part of the universe we see today. Mathematical logic presents us with a fait accompli. There must exist a singularity in our past, a boundary to time from which the expanding universe emerged. The proof is a difficult piece of abstract mathematics, in itself a remarkable testimony to the ability of such reasoning to produce and use statements about physical reality. But despite the difficulty, the essence of why the presence of a trapped region leads to an inevitable hole in the fabric of space and time reduces to a single simple idea. Light originally emanates from a region with the configuration of a flat, open sheet, but once a trapped surface forms around it, the region takes on the configuration of a ball. Now, the only way a smooth distortion of a plane into a spherical configuration can be effected is if at least one point is missing from it. This is why projections of the surface of the earth from a spherical globe onto a flat atlas have to be distorted. A continuous journey around the spherical globe would be broken when traced on the flat Mercator projection in an atlas. In the case of the universe, there must also be missing points within the trapped region that is our past. These missing points, and there could be an infinite number of them, mark the singular boundary that we call the beginning of the universe.

White holes

One of the most unusual features about the proof establishing our new concept of a singularity is that it is by no means necessary that everything in the universe experiences or issues from that singularity. In fact, just one single point will do. Large portions of the universe might have been excluded from this fate. Also, the singularity almost certainly will not be simultaneous. Different points of the universe may begin their expansion from a "big bang" at different times. The denser portions of the universe, which eventually condense prematurely into galaxies, probably emerged from the "big bang" slightly later than those regions that now compose the sparse intergalactic medium. The existence of small nonuniformities in the universe implies that creation could not have been simultaneous. More striking still, some portions of the singularity could have been so delayed at birth that their creation could have been influenced by other parts already in existence.

In Figure 2.3, light signals emanating from *A* could influence events at *B*. These are dramatically retarded pieces of the "big bang" and are sometimes called white holes. If we lived close to *B*, then we would experience entirely unpredictable local phenomena from the singularity on our doorstep, despite observations of the universal expansion that made us think the real "big bang" singularity lay in our far distant past at *A* (see Figure 2.3). A white hole would appear to us as a black hole in reverse (hence its name), inexorably expelling matter as if from nowhere and in violation of all the laws of energy conservation. At one time such exotic creations were considered as viable explanations for spectacular cosmic energy sources like *quasars*. Although no larger than the solar system, quasars emit more energy than an entire galaxy of one hundred billion stars. However such an association proved difficult to substantiate when more detailed facts about quasars became known. There is now believed to be no positive observational evidence for the existence of white holes.

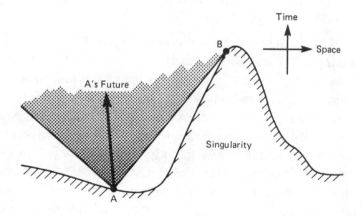

Figure 2.3   *Non-simultaneous Creation*. Light signals emitted at the creation of *A* can influence events at the creation of *B* if *B* lies in *A*'s future.

One of the most unusual things about these inevitable singularities is that, superficially at least, they are starting points to cosmic history that need not necessarily go hand-in-hand with the infinite densities and other cosmic fireworks of a "big bang." So far, a lot of effort has been expended in trying to prove that such infinities really do accompany a singularity. If this were true, it would tell us that, in these cases, it is the infinite densities that are the reason for the universe possessing an edge in time. A number of very technical and elaborate conclusions have been drawn so far concerning this question. They are best summed up by saying that it appears that the majority of realistic singular universe models do possess infinities at their beginnings, and the particular model universe that best describes our own today is probably (but not definitely!) one of that number.

Did the universe begin with a "bang" or a "whimper"?

"This is the way the world ends. Not with a bang but a whimper." These words are taken from T.S. Eliot's poem "The Hollow Men." The term "big bang" was first coined by Fred Hoyle in a radio broadcast on cosmology in 1950 and is usually reserved as a description of the dramatic type of singularity at the start of the universe where physical quantities become infinite. More recently in 1974, George Ellis and Andrew King have used the label "whimper" for a more moderate boundary point of the universe where physical quantities like density and temperature all retain moderate values. Several studies have been made in an attempt to ascertain whether the beginning of our universe was a "bang" or a "whimper." It appears to be far more likely that the expansion began from a "bang" rather than from a "whimper." There are far more possible models with a "bang" than with a "whimper" contained in Einstein's theory of gravitation. More significant though, is the fact that if we change those with a "whimper" by a tiny amount, they turn into "bangs"; whereas, if we change those with "bangs" very slightly, they remain "bangs." "Whimpers" are unstable; they require very, very special cosmic starting conditions, whereas "bangs" seem able to occur under many kinds of conditions.

One reaction to the inevitability of the "big bang" and its initial singularity has been that an alternative may exist because no known theory of gravity will apply when the density gets arbitrarily large. There are two things to be said about this. Firstly, the theory should only break down when the density reaches 10^{96} times that of water, and such a fantastic density is already pretty singular by anybody's standards. Secondly, the possibility of "whimpers" as a consequence of the theory cannot be swept under the carpet by this objection. There appears no reason why the theory should break down on approach to a "whimper" because conditions are very moderate there. In these cases, the prediction of general relativity, that it leads to a state in which it cannot make any predictions, an edge to space and time, is inescapable. General relativity, as was said of something else more political, "contains within itself the seeds of its own destruction."

The most difficult question to answer about a singularity, and one which still is largely unanswered, concerns what goes on in its immediate vicinity. What would the universe look like in the neighborhood of a typical singularity? And, if there really was a singularity at the beginning of our universe, was it of the typical, run-of-the-mill sort or could it have been special in some way that is still reflected in some of the peculiar properties of the present-day universe? In chapter 5, we shall return to ponder some of these questions because they play such a crucial role in evaluating whether the present cosmic structure evolved out of primordial order or from primordial chaos.

Region of no escape

The bizarre singular points that mark out the edge of space and time seem at first remote from our present experience of the universe. But a *local* trapped region with a singularity lurking within may exist at the center of our own galaxy or in other galaxies. These singularities are not part of the "big bang" singularity, but they are the holes that must appear locally in any region that contains enough material at high density to become trapped. This can happen quite easily in the dense regions at the center of a galaxy. Astronomers call these local trapped regions black holes. The reason for their existence can be appreciated in very simple terms. Again, gravity alone is the villain of the piece.

Throwing a stone is a familiar manifestation of gravitational attraction. After tracing out a roughly parabolic trajectory, the stone hits the ground. The harder we throw the stone, the higher it goes and the longer we must wait before it returns to earth. This suggests to us that ultimately we might be able to overcome gravity. Perhaps, if thrown fast enough, an object could escape completely from the pull of the earth's gravity and disappear into space, never to return. This indeed is possible. Everything has associated with it a critical speed, often called escape velocity, the speed which another object must attain to escape completely from the gravitational pull of the earth or parent body. This velocity is determined by the mass and radius of the larger body together with Newton's gravitational constant. Newton's constant specifies the intrinsic strength of gravity. For the earth, these quantities combine to determine the speed that must be achieved to escape and yield an escape speed of 11 kilometers per second or 25,000 miles per hour. This is the minimum launch speed that must be attained by a rocket if it is to escape the earth's gravity. The concept of escape velocity is useful in many applications. It enables us to understand, for example, why the earth retains a gaseous atmosphere while the moon does not. Close to the earth's surface, the average speed of oxygen and nitrogen molecules is less than the terrestrial escape velocity so they are retained. However, the moon has 1/81 the mass and 27 percent the radius of the earth. The moon's escape velocity is nearly five times smaller than the earth's, and gas molecules easily escape into space from its surface. Gravity is what determines whether planets have atmospheres.

Black holes conceived

John Michell was an English clergyman and geologist who became famous for his invention of the torsion balance and the creation of seismology as a science. In 1783, his attentions were focused on the problem of gravity. Intrigued by a very unusual idea, he wrote of it to Henry Cavendish, a famous Cambridge physicist of his day. Michell knew all about escape velocities, and he guessed that light was composed of tiny corpuscles moving at high velocity (we would call them "photons" today). He asked himself what would happen if an object was very massive, yet small enough for gravity to be so intense that it possessed an escape velocity equal to the velocity at which light moves (300,000 kilometers per second). Such an object, Michell surmised, would be invisible to distant observers. None of the light falling upon it from outside could be reflected and escape to a distant observer's eye or telescope. Light would be trapped by the object. Michell calculated that an astronomical body with this bizarre property would have to be about ten million times heavier than the sun in order to have the same density as the earth. He even suggested that the presence of one of these bodies might be detected if it were in orbit around another visible star. Any anomalous orbital motion of the visible star would imply the existence of its unseen companion.

Michell had predicted the existence of what are now called black holes, a name coined by the American physicist John Wheeler in 1968. Several astrophysical objects are now prime candidates for being black holes.

Black holes observed

Cygnus X-1, an otherwise nondescript binary star in the constellation of Cygnus and an intense and highly variable source of x-rays, is suspected of being a black hole. The x-ray emission is the prime clue. This emission requires the release of gravitational energy by gas as it falls onto an exceedingly compact object, just as an object falling onto the ground from a considerable height acquires energy which is released on impact in the form of heat. Neutron stars and black holes offer the only two possibilities for the compact object. The mass of Cygnus X-1 can be estimated by studying the orbit of its ordinary companion star spectroscopically. This star is estimated to be in excess of five solar masses. Neutron stars are known to be unstable if they weigh more than three or four solar masses—collapse to a black hole is the inevitable fate for them (see page 150). Hence, the conclusion arrived at by astronomers. Cygnus X-1 is almost certainly a black hole. There also exist other x-ray sources that astronomers suspect are black holes, but none offers such compelling evidence as this (see Figure 2.4). [2]

Much more massive black holes are inferred to be present in the nuclei of active galaxies. Material falling under gravity onto such objects offers the most plausible and efficient

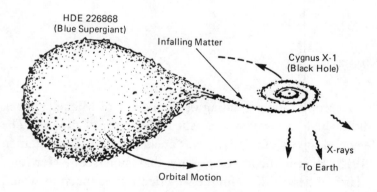

Figure 2.4 *Cygnus X-1.* The x-ray source Cygnus X-1 is capturing material from a supergiant star. As this material spirals into the black hole, it forms a disk, heats up, and emits x-rays.

means of extracting large amounts of energy from matter in a compact region. In an extreme situation, such as in the quasar 3C 273, a luminosity equivalent to one hundred thousand Milky Ways is produced within a volume smaller than a cubic light year. Only for nearby galaxies do astronomers have sufficient telescopic resolution to identify optical evidence for the presence of large black holes. One favorite hunting ground has been the nucleus of the nearby giant elliptical galaxy Messier 87. There are tantalizing indications that the central star density is abnormally high and that stellar velocities are also unexpectedly large—precisely what one would expect to see close to a large black hole of several hundred millions of solar masses. Better optical resolution is only attainable with access to a large telescope in space. When that development occurs in 1986, astronomers will seek to confirm that the innermost stars in galactic nuclei are responding to the gravitational pull of a compact and very massive object, and that object will almost certainly be a black hole (see Figure 2.5).

Event horizon

A proper understanding of the nature of Michell's black holes had to await a sophisticated theory of gravity. The general theory of relativity was created by Albert Einstein in 1915. It is a theory of gravity that supersedes Newton's classical description—in the way that a time-machine would supersede the first flying machines. Newton's theory was developed, and is perfectly adequate, for the weak gravitational fields in the solar system. Einstein's theory opened up

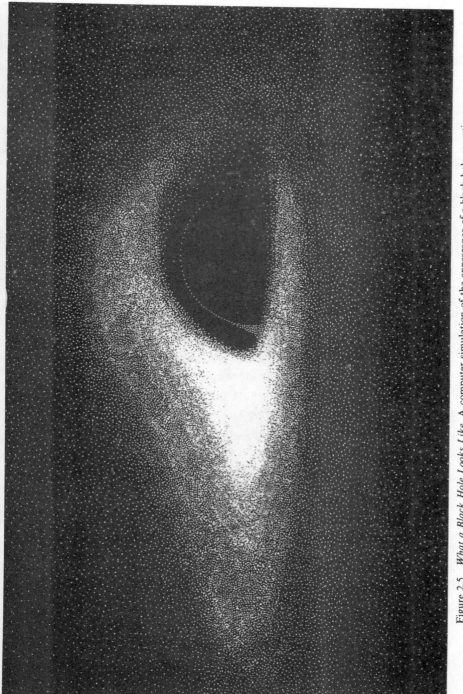

Figure 2.5 *What a Black Hole Looks Like.* A computer simulation of the appearance of a black hole accreting gas.

entirely new vistas that explain intense gravity fields. Unlike Newton's theory, it accurately predicts the behavior of objects moving at speeds approaching that of light, and describes the motion of matter in regions of very strong gravitational attraction, even near black holes. Whereas Newton's theory is mathematically simple, just a single equation determining the forces exerted by a given distribution of masses, Einstein's theory is extremely complicated. Instead of one, we now have ten intricately interwoven equations that relate the structure of space and time to the distribution and motion of matter within them. Even now, after over sixty years of detailed study, the equations have only been solved when special symmetries or approximations are assumed. One of the first, simplest, and most important solutions to Einstein's equations was discovered by the astronomer Karl Schwarzschild in 1916. After many years of debate, it became clear that Schwarzschild's solution was the precise, general relativistic description of the black hole state envisaged by Michell. It is spherical, and once we know its mass, we know everything there is to know about it. In particular, we know the radius inside which the gravitational field is strong enough to prevent the escape of anything, even light. The radius, given by Michell's formula,* corresponds to an escape velocity equal to the speed of light. The spherical surface of this radius is called the *event horizon* of the black hole. No signal can be transmitted across the event horizon from the interior to the external world. It is a one-way membrane. Once a spaceship has crossed the event horizon and entered the black hole, it cannot escape back out no matter how powerful its rocket engines. To give some feel for the quantities involved, if our galaxy had a black hole with mass equal to a million suns

---

*$r = 2GM/c^2$; G is Newton's gravitation constant; c, the speed of light; r and M, the radius and mass of the body.

lurking at its center (which is not only possible but quite likely), then the event horizon would be a spherical surface with a radius about one fiftieth of the distance from the earth to the sun. If the sun (with a mass 1.99×10^{33} gm) were to collapse to a black hole, its radius would become a mere three kilometers, more than 200,000 times smaller than at present.

From the outside, black holes are the simplest objects in the universe. There are only three properties which they possess that an outside observer can ascertain. These are mass, angular momentum to describe any rotation, and electric charge. Schwarzschild's solution was the simplest case possessing zero angular momentum and no charge. The general solution possessing all three features simultaneously was found in 1966 by Roy Kerr, a New Zealand mathematician. These three qualities that a black hole can possess are those believed to be absolutely conserved in nature. Each can be redistributed in a physical process, but can never be absolutely destroyed or created. Black holes possess no other properties. Two black holes with identical mass, charge, and angular momentum could not be distinguished in any way. They carry no other memory of the material from which they were formed. No outside observer could tell whether it was matter or antimatter, solid or liquid, radiation or particles, mice or men, that had collapsed to form the black hole. This result is known to physicists as the "No Hair" theorem. By saying black holes have no hair, we mean simply that they possess no superficial properties that make them individually distinguishable when their masses, charges, and angular momenta are equal. This does not mean that matter falling into the black hole ceases to have any other properties, only that these three are the only ones that can be discerned by someone on the outside. All other outgoing information gets blocked by the event horizon.

Naked singularity

Black holes clearly create very unusual large-scale effects in the universe, but strangely, they need not bring about any unusual local events. We could all be living inside a black hole now without noticing anything amiss. The average density of matter contained within the large black hole at the galactic center that we previously hypothesized would be only that of air. If we crossed its event horizon at this very moment, we would notice nothing unusual, no enormous forces or weird phenomena. But if we tried to reverse our path and backtrack across the event horizon, we would discover that we could not return. Once behind the horizon, furthermore, we would eventually find ourselves dragged inexorably towards the center of the black hole. As we approached the center, we would experience stronger and stronger gravitational forces. These forces would finally tear our spaceship, and our bodies and constituent atoms, to pieces.

Suppose instead we were a light ray. What would happen if we plunged into the center? As we have seen, Einstein's theory, and other theories like it, predict that we would eventually encounter a hole in space and time, a singularity, where space and time come to an end. It is not possible to pass through this point; it is part of the boundary of the universe, just as a "big bang" singularity would be. If we encountered it, we would simply cease to exist because the very space and time in which all being is grounded are destroyed. The reason for this singularity is what one might expect—the infinite strength of the gravitational forces there. Anyone passing toward the center through the event horizon has an unavoidable encounter with the singularity within a finite time. These predictions were first made by the Oxford mathematician, Roger Penrose, in 1965. He proved that a hole must develop within any local region of the universe trapped by a horizon.

There is one last aspect of the black hole singularity that is intriguing. Because the hole is shrouded by the event horizon, no information about events near it or information issuing from it can escape to us outside the event horizon. Penrose's proof predicts that the laws of physics, as we know them, must break down at the black hole singularity. Anything can happen there. We cannot predict events, and all predictive science is impossible in its vicinity if material is emitted from the singularity. However, if whenever a singularity arises in the universe (and there could be, at the very least, one at the center of every galaxy and quasar), it is always shrouded by an event horizon, then the irrational, unpredictable phenomena would be proscribed and unable to influence the observations of external observers like ourselves. Penrose calls this hypothesis cosmic censorship. If it is false, then locally we could encounter "naked" singularities unclothed by event horizons. What would one of these naked singularities look like if we could observe it?

There seem to be no rules at all about what could happen near a naked singularity. At first, this creates a picture of complete chaos, with the unfortunate neighbors of the naked singularity finding all their long-cherished laws and regularities of nature completely overturned. While this is quite possible, it is by no means compulsory. Partial randomness is impossible to predict, but complete randomness is something with which physicists are quite comfortable.

Completely random systems seem to obey predictable statistical laws. Although an individual event cannot be predicted, the trend of an on-going sequence of events can.

Indeed, this is the situation in the description of the air motions in the room at any moment. The motion of a single molecule cannot be predicted because of the complexity of its innumerable collisions with the others. But, because those collisions are so numerous and complex, they are random by virtue of being completely unbiased. Despite this microscopic indeterminacy, the gross motion of a large body of air is predictable. Naked singularities might turn out to be no more unpredictable than any other statistical system with built-in random noise such as a television set. This brings us full circle back to cosmology, for there certainly does appear to exist at least one naked singularity: the "big bang." Our observations of the universe show that it seems to have been highly ordered rather than chaotic. The earth, the stars, and the galaxies are just a few of its outpourings.

The chain of deductions we have followed to conclude that our universe was singular in its past is the result of a *theorem* rather than a *theory*. Certain precise assumptions are made and from them, the conclusion drawn is based on mathematical logic alone. Although the assumptions are couched in mathematical language, they are seen to hold true in the universe today. But what about in the distant past? If the singularity really did involve infinite densities, surely, as we approach it, all sorts of new physical phenomena might arise. New laws and new forces might violate the simple assumptions that we have drawn from our present experience of the world.

Some scientists find the conclusion that a singularity in Einstein's theory is inevitable to be highly distasteful. That it cannot predict before a certain time in the past may be a signal that something has gone wrong with Einstein's theory rather than with the real world it purports to describe. If we wish to evade the conclusions of the singularity theorems, it is necessary to pick a quarrel with at least one of their basic assumptions.

Suppose it were believed to be more natural that the universe "bounced" into expansion when it was squeezed into a tiny volume. How could we support such a view? The assumption that has been most consistently attacked as the weak link in the chain of logic leading to the space-time singularity is that which assumes gravity is always attractive. The thrust of this attack, which is also an attack on the validity of general relativity, is the fact that Einstein's theory cannot be the ultimate description of the world; in particular it is not a quantum theory. This fact alone guarantees it cannot be the ultimate description of a universe of immense density and temperature. But, then again, do we know whether quantum theory could save us from the singularity?

Quantum uncertainty

The underlying idea of a quantum theory is that particles and energy fields exhibit a dual character. They possess behavioral traits of both waves and particles. They can, on occasion, be made to behave like colliding billiard balls or equally, as interfering and oscillating waves. This dual relationship is extremely subtle, and it is most helpful to think of the wave aspect as a means of carrying information about a particle's location and motion. The idea of a crime wave is a more appropriate analogy than a water wave. If a crime wave hits an area, it is a statement about probability. It tells us that we are more likely to find a crime committed there than outside the crime wave area. The other hallmark of a quantum theory, which springs from the existence of wave-particle duality, is the *uncertainty principle* proved by Werner Heisenberg. This says it is impossible to measure precisely both the position and the velocity of an elementary particle at the same time. An uncertainty must always exist in the product of the ambiguity in a particle's position, $\triangle x$, and the ambiguity in its momentum (mass times velocity), $\triangle p$. The smallest value of this combination that can be measured is given by one of the fundamental constants of quantum mechanics, Planck's constant divided by $4\pi = 12.57$. This is inferred from the equation:

$$\triangle x \, \triangle p \geq \frac{h}{4\pi}$$

The numerical value of h is very small (6.625×10^{-27} erg second) and for large, everyday objects, the uncertainties in position and momentum are negligible. It is important to appreciate that the uncertainty principle has nothing to do with any fallibility of our measuring devices or limitations in our models. The uncertainty it predicts is irreducible even by perfect measuring instruments. Its origin is simple to see.

Suppose we knew the momentum and hence the exact speed of a microscopic particle. For example, take it to be at rest so the momentum is precisely zero. Could we now know exactly where the particle was in space? The uncertainty principle says no, and the reason is that when we try to measure its position, we inevitably move it in the process. By bouncing a photon off the particle into our microscope, we change its position. By the time we "see" it and record its position in space, it is no longer at that initial location. The uncertainty principle tells us that the uncertainty that the act of measurement creates is at least as big as Planck's constant. For large objects, like a football or an automobile, these perturbations created by the process of observation are infinitesimal and completely irrelevant in practice. But in the world of elementary particles, they are vitally important.

Quantum gravity

This inevitable complementary uncertainty in the motion and position of a particle illustrates why it is so difficult to make Einstein's theory of gravity into a quantum theory. Einstein's general relativity has unusual features rivaling any of the aspects of quantum theory. Usually, theories of nature provide us with sets of rules, in the form of equations, for predicting how objects will move and interact with one another once we have placed them in some previously chosen geometrical space. Newton's laws of motion are like this. Among other things, they will predict the behavior of two billiard balls colliding on a table. But general relativity is far more sophisticated. In this theory, it is quite unnecessary to specify Euclid's axioms of geometry as Newton would have done, or indeed presume anything about the spatial geometry in which particles are placed. Their very mass and motion determine the geometrical fabric of the space and time in which the particles move, rather as ball-bearings rolling on a rubber sheet would completely determine the local topography of the sheet as they moved about on it. The contrast with the old Newtonian picture is illustrated by a little parable:

Suppose a population of glowworms set out one night to

take the quickest route between two points. As we watched them from above, we would witness a single procession of glowing points moving in a straight line. All of a sudden, the lights all veer along a curved path before resuming their straight-line route. A Newtonian physicist would say, all bodies acted on by no forces must continue to move in a straight line or stay at rest; since the glowworms have deviated from their straight path, some force must be acting upon them. The Einsteinian physicist turns on the light and points out that there is a ditch in the path of the glowworms. They simply took the quickest route around the surface of the ditch. When the surface of the ground ceased to be flat, this was the straightest path they could take, just as it is for aircraft that fly great circle routes over the earth's surface.

Einstein showed how the curvature of space explains the idea of mysterious forces acting between distant bodies. When a mass is placed in space, it curves the space in its vicinity like an object placed on a rubber sheet. So, when another particle comes nearby, it rolls towards the first mass, not because there are any long-range forces present, but because it responds to the local curvature of space created by the first mass and takes the quickest route between points— just as the glowworms did.

The clash of quantum and relativistic concepts can now be clearly seen. For the uncertainty principle forbids us from knowing the exact position and motion of any particle when knowledge of these facts is precisely what is necessary to specify the geometrical structure of space and time in which they reside. If so, then we must admit that the whole space-time structure of the universe is indeterminate. But, then, how can we devise a prescription for introducing our particles in the first place? We are caught in a vicious circle of inconsistency!

It is evident that general relativity must be drastically modified by quantum principles and quantum theory extended to include the effects of curved space and gravity. This is why the assumption made by the singularity theorems, that gravity is always attractive, must be strongly questioned as one gets close to the physical extremes of a singular state. These objections may not save us from the existence of a singularity though, for as we have stressed above, singularities need not always involve the extremes of density and temperature that could invalidate the assumptions of the theorems. Indeed, there could be a space-time singularity on this page, smaller than an elementary particle and emitting random information into the universe. We would notice nothing bizarre and would be unable to discriminate from the randomness that is generated simply by collisions between atoms and molecules. However it appears likely, as we will see later, that the singularity in our universe's past did involve extreme conditions. So with that assumption we can ask just how close to a singularity our theory allows us to get before it fails.

Planck moment

At what moment do the unusual effects of quantum theory, ignored by Einstein's theory of gravity, become overwhelmingly important? Fortunately, nature tells us the answer. In a universe governed partially by the influence of gravity, light propagation, and a quantum theory of matter, there exists a unique time at which all these three effects are of equal importance. The strength of gravity is described by

Newton's constant of nature, $G = 6.672 \times 10^{-8} \text{ cm}^3 \text{ gm}^{-1}$ sec^{-2}, the uncertainty of quantum theory via Planck's constant, $h = 6.625 \times 10^{-27} \text{ gm cm}^2 \text{ sec}^{-1}$, and the relativistic theory of light characterized by the speed of light, $c = 3.0 \times 10^{10} \text{ cm sec}^{-1}$. If we look at the units of mass (gm), length (cm) and time (sec) we have chosen, we see that it is possible to combine these three constants of nature and create a mixture with the units of a time in one and only one way. Take G, multiply it by h, then divide by c five times, now take the square root of your answer; notice that the units of your answer are seconds and its magnitude, t_p, is

$$t_p = \sqrt{\frac{Gh}{c^5}} = 1.33 \times 10^{-43} \text{ sec}$$

This is undoubtedly the shortest interval of time the reader has ever encountered. To put it in perspective, the time it takes for light to cross an atomic nucleus, 10^{-24} second, is huge by comparison. The time t_p, although first discovered by the Irish physicist, George Johnstone-Stoney,[3] in the 1870s, is called the Planck time after Max Planck, one of the pioneers of quantum theory who discovered it independently in 1906. It is a fundamental time picked out by the way nature is woven together and is independent of any artifacts or human clocks. If we wanted to explain to someone on a distant galaxy what the average life expectancy of a human was, we could tell him to measure the strength of gravity to find G, the velocity of light, and the quantum of uncertainty in his laboratory. He would then evaluate them in his own peculiar, extra-terrestrial units of arithmetic, length, mass, and time. Then we could give him the prescription we gave the reader for constructing t_p and tell him that humans live for about 1.66×10^{52} Planck times, ($= 70$ terrestrial years; note 1 year $= 3.156 \times 10^7$ secs).

Origins

New beginning

The Planck time also signals the breakdown of our present theories. Before 10^{-43} second, we have no good idea as to the nature of the universe. A major extension of physics will be necessary to delve into the first 10^{-43} second and follow the history of the universe in those first quantum moments when it was entirely shrouded in uncertainty. At the moment, when cosmologists talk about the "beginning of the universe," they are using a form of verbal shorthand. When pressed, what they usually mean is the Planck moment. This is the time after which serious and reliable mathematical calculations can be framed with some confidence in the underlying physical theories. For all practical purposes, it is the beginning of the universe. At the Planck moment, the temperature of the universe is predicted to be 10^{32} degrees Kelvin and the density to be 10^{96} times higher than that of water. So we see that in order to avoid the conclusion that a singularity exists, we have to enter an environment that by anybody's standard is pretty singular!

These quantum problems tell us that things might go wrong with our predictions when we try to extrapolate them to times earlier than 10^{-43} second. But what sort of things might go wrong? During the last few years cosmologists have begun to make brave attempts at predicting events very close to the Planck epoch. They have not yet been able to accommodate a quantum uncertainty in the space-time fabric but have made exciting progress in elucidating how quantum particles and waves might behave in a conventional model of time and space. This tractable problem has produced results that alter our picture of the Planck moment and may, in the future, yield some experimentally testable consequences. [4]

Quantum vacuum

One of the naive notions that quantum theory overturns is our image of a vacuum. We are used to imagining a vacuum as empty, pure and simple. But now we see that the uncertainty principle forbids us from making any such statement. Because we can never observe a vacuum without introducing particles into it, we can never say what its *precise* content is. What we have to do in order to be consistent is to picture the vacuum as a sea of continuous activity, a vast foam of continuously appearing and disappearing pairs of oppositely charged particles (see Figure 2.6). Each of these pairs is *unobservable* because the distances they move between creation and annihilation, and their momenta, p, satisfy (*), Heisenberg's uncertainty principle condition. So, according to this principle, they are unobservable. Such pairs

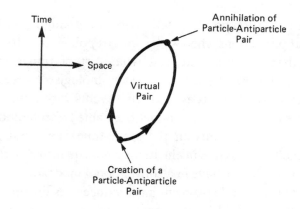

Figure 2.6 *Virtual Particle Pairs.*

are, therefore, called virtual particles. Their creation from "nothing" involves a violation of energy conservation, but nature does not mind this so long as it is unobservable. In order to make our transient virtual particles, energy must be "borrowed," and the uncertainty principle is nature's way of underwriting such a loan. The loan can be as large as you like, but the larger it is, the shorter the time allowed before repayment must be made through annihilation. This prevents the individual loan being observed. Another, equivalent way of writing the equation on page 58 is

$$\text{(magnitude of the energy loan)} \times \text{(period of loan)} \geq \frac{h}{4\pi}$$

The reader is probably beginning to feel sucked into mysticism. Why should we bother with such an absurdly elaborate construction that common sense dictates ought to be simplicity itself. And worse than that, we seem to be saying the whole elaborate charade is unobservable anyway, so why bother! But while individual virtual pairs cannot be observed, we can observe consequences of the entire sea of them. At any particular instant, there will be many virtual pairs caught between birth and annihilation. They are predicted to have a calculable effect upon the energy levels of atoms. The effect expected is minute—only a change of one part in a billion, but it has been confirmed by experimenters.

In 1953 Willis Lamb measured this excited energy state for a hydrogen atom. This is now called the Lamb shift. The energy difference predicted by the effects of the vacuum on atoms is so small that it is only detectable as a transition at microwave frequencies. The precision of microwave measurements is so great that Lamb was able to measure the shift to five significant figures. He subsequently received the Nobel Prize for his work. No doubt remains that virtual particles are really there.

Negative energy

The presence of virtual particles creates unusual physical effects. The most striking is a phenomenon originally predicted in 1948 by the Dutch physicist Hendrik Casimir. If we cool a physical system down to a very low temperature, all the noise and thermal motion will eventually cease, leaving only the activity of the virtual particles demanded by the uncertainty principle. This motion is sometimes called the "zero point" motion. It has a wave-like character and is a superposition of waves of all possible wavelengths. Casimir asked a very simple question: what would happen to the vacuum fluctuations if two parallel plates were placed in their midst?

Outside the plates all possible wavelength fluctuations can exist, but between the plates we can only have those waves which can exactly fit a whole number of wavelengths between the plates. Clearly, there must be fewer waves between the plates than outside them, and so there must be more pressure on the outside of the plates than the inside. The plates must, therefore, move together. If the plates are one ten-millionth of a centimeter apart, the force pushing them together is about 130 dynes. The pressure they feel is roughly ten thousand times smaller than atmospheric pressure at the earth's surface. This tiny pressure was measured in 1958.

What the Casimir effect is telling us—besides providing

more experimental confirmation of our picture of the quantum vacuum—is that negative energy densities can exist. If we take a quantum vacuum, it is, by definition, the state of zero energy. When the Casimir plates are added to it, some virtual particles can no longer be accommodated; therefore, the energy density must be smaller than zero, and so the energy density is negative. At the Planck time, particles of matter or gravitational fields can play the role of the Casimir plates. Regions of negative density are created that lead to gravitational *repulsion* rather than attraction. By this means, cosmologists hope that the attractive nature of gravity might be reversed and the conclusions of the singularity theorems evaded.

Creation

The last and most exciting phenomenon that is expected to occur at the Planck moment is the spontaneous creation of *real* (rather than just virtual) particles. In 1951, Julian Schwinger predicted that if a high voltage was applied across the Casimir plates, so creating a strong electric field, then virtual particles can be turned into real, detectable particles. If the electric field is very strong, the voltage will vary so much from place to place that when the components of a virtual pair separate in space, they will each feel significantly different electric forces upon them. This can so alter their trajectories that they subsequently fail to encounter one another again and so do not annihilate. They become real and observable. No observable violation of energy conservation occurs. The energy budget is balanced because the energy needed to create the real particles is taken from the energy of the electric field. This process has also been observed in the laboratory. A similar effect is expected to occur at the Planck epoch of the universe, not due to an electric field but due to the intense gravitational field. Virtual pairs will appear spontaneously and then, soon afterwards, annihilate. But near the Planck moment they appear in a medium of enormous density with a super-strong gravitational field that varies dramatically from place to place. When pairs appear, each member will feel a slightly different gravitational force in its vicinity. This prevents them from annihilating again. Instead, the pairs become real at the expense of the nonuniformity in the universe's gravity field. It is possible, in fact, that this process of particle production is the reason why the universe is so smooth on the average; the creation of particles may have ironed out lumps and bumps at the Planck moment.

Black hole evaporation

In 1974, Stephen Hawking showed that we might be able to observe gravitational particle production in space. He calculated that this same process of virtual pair creation occurs at the boundary, (that is, the event horizon), of a black hole. If one member of a virtual pair falls outside of the event horizon, it will not annihilate. Rather, the particle will escape into space as a real particle, leaving the black hole very slightly less massive than before (see Figure 2.7).

The more curved the black hole horizon, the more pronounced the effect will be. Since the radius of a black hole

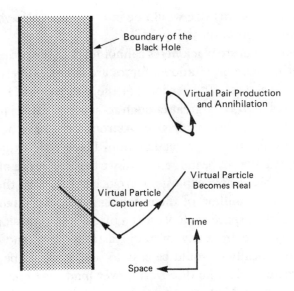

Figure 2.7 *Virtual Particles Near a Black Hole.*

is proportional to its mass, the smallest and least massive black holes have the most strongly curved horizons. The long-term effect of Hawking's process is rather startling: *the black hole will slowly evaporate.* Quantum black holes are not completely "black." As the process continues, the black hole's mass falls, and the curvature of the boundary becomes stronger as the particle creation goes faster. Eventually, the entire black hole disappears in an explosion leaving behind (perhaps!?) a naked space-time singularity at its center. Hawking showed that black holes with a mass of about 10^{14} gm, about the mass of a small mountain, which have a diameter on the order of the atomic nucleus, 10^{-13} cm, take fifteen billion years to evaporate—about the age of the universe. So black holes of this size would be in the final explosive phase of their evaporation today. [5]

Such miniature black holes cannot form in the present-day universe where gravitational forces are comparatively weak. Only in the first 10^{-23} second of the universe's life are the pressures of gravity great enough to squeeze material into a black hole of this small size. Astronomers have searched carefully for these spectacular mini black hole explosions, but none have yet been found. All we can say at the moment is that if black hole explosions are occurring, then there are fewer than a million of them happening in every ten cubic light years of space each year. If a black hole explosion were ever observed by x-ray or radio astronomers (and if one occurred locally it would be easy to see), it would be one of the greatest scientific discoveries ever made. At one instant of time, we could observe a quantum gravitational phenomenon that is essentially the "big bang" in miniature and which leaves its space-time singularity visible. We would be able to examine a cosmic fossil that has survived from the first 10^{-23} second of the Universe's life.

Cyclic universes

As the offspring of a marriage between the theories of quantum mechanics and gravity, these exotic new physical processes and principles offer the possibility that the "big bang" singularity can be subtly avoided. It is equally possible, of course, that the "big bang" singularity's existence might be preserved or even reinforced by the correct quantum model. The first naive quantum cosmological models have nothing definite to tell us yet. Some avoid a singularity, some do not. Gradually, as knowledge and ingenuity develops, we expect the number of self-consistent possibilities will be whittled down. Of the existing possibilities, the one that the reader most naturally thinks of is probably that of a closed cyclic universe oscillating through an infinite sequence of recurring expansions and contractions (see Figure 2.8).

This model is not feasible without quantum cosmology. The singularity theorems absolutely forbid anyone or anything from re-emerging out of a singularity, just as escape from a black hole is impossible. However, any particles that do not hit the singularity could pass from a contracting universe to an expanding one unscathed. Suppose that an oscillating universe model were feasible. It is not clear whether we could be living in the latest of an infinite sequence of oscillations stretching back to past infinity and on into eternity. Attractive though this idea is, it is spoilt because, unfortunately, as the oscillations proceed, radiation tends to accumulate. The accompanying increase in pressure tends to make each cycle a little bigger than the last. The expansion will also build up more and more chaotic ir-

Figure 2.8 *Cyclic Universe.* The oscillating universe has each cycle slightly larger than the preceding one owing to the steady increase of the heat content of the universe.

regularity as the cycles continue. We do not seem to live in a universe that has inherited such an unusual structure from its predecessors, but we really do not have anything but speculation to go on. Suffice it to say that all ideas about bouncing, cyclic universes are likely to remain in the realm of science fiction for some considerable time to come. But sometimes fact is stranger than fiction.

3

Creation

THE early universe was a paradise for the elementary particle physicist, but it is a paradise lost. The universe has expanded and cooled. The symmetries of high energy are broken and disguised, and the exotic particles are gone, leaving only here and there a clue to the past—a past unimaginably different from the present.

Forces of nature

Today the world appears to exhibit a vast panoply of forces. The pull of terrestrial gravity is the most familiar, but there are also magnetic forces of attraction and repulsion, adhesive forces and frictional resistances, electrostatic forces, explosive and muscular forces. All of these, to a greater or lesser extent, are part of everyday life. We know also of the forces of celestial gravity that create the tides and control the orbits of the earth, the moon, and the other distant planetary bodies. Finally, there are the powerful nuclear forces that create both vast energy sources, weapons of destruction, and ironically, solar power. As human beings, we feel only some of the overt consequences of these forces. Their effects differ so greatly that we are prejudiced into believing them to be also quite distinct in origin. Magnetism does not seem to bear any superficial resemblance to light, nor does electricity seem to have any link with nuclear explosions.

It has always been the goal of physicists to discover the simplest possible explanation for the forces of nature in all their various manifestations. In the course of these detailed investigations, the number of distinct types of force field has been steadily whittled down.

Until about fifteen years ago, the *weak, electromagnetic, strong,* and *gravitational* forces appeared to physicists as unalike as chalk and cheese. They possessed among them all that was necessary for us to make sense of the visible world. Each superficial force we encounter among the vast range on earth and in space is a disguised example of one of these four fundamental forces. Yet this quartet appears different in virtually every respect: each force possesses entirely different strength and range. Each force influences different sets of elementary particles, with the honorable exception of gravity which is unique in that it appears to act on everything. With evidence such as this, there seems little hope for the type of simplification that great physicists like Einstein and Eddington had sought—that of a completely unified theory of fundamental forces. Einstein devoted the last forty years of his life searching for a description of gravity and electromagnetism that would reveal them to be just different manifestations of a single more basic interaction. His search led nowhere. We now know with the benefit of considerable hindsight, that Einstein made the worst possible choice of candidates in his pursuit of unification.

Gravity is a most unusual force. It stands apart from the other three forces of nature in that no means has yet been found to apply the quantum theory to it nor it to the quantum theory. The *strong, electromagnetic,* and the *weak* forces are more akin, and are all described by quantum theories. It has been through their detailed study that exciting glimpses of unity in the face of subatomic diversity have recently begun to emerge. It has been realized that nature has disguised the unity between these three forces, and maybe even between all four, in a special way, by making their strengths dependent on the environment in which they are measured. This does not mean that the fundamental constants telling us the intrinsic strength of these interactions of nature are varying in space or time. Rather, the strengths of their measured *effects* vary with the temperature at which they are measured.

Human beings are built from fragile DNA molecules that can only exist in a relatively cool world, and so in our usual experience, we see only the way the fundamental forces behave at the relatively low energy where biology is possible. If, as we shall see, we could have witnessed the extremes of energy and temperature at the time of the origin of the universe, we would have deduced something quite different about the ultimate structure of the forces of nature. What exactly would be different? If we do some simple thought experiments, we can unveil a little of nature's cunning disguise.

The Left Hand of Creation

Vacuum

The electromagnetic force strength is determined by the electric charge of an electron, the basic unit of electric charge. Imagine two electron charges, A and B, of the same sign approaching one another. In a world without quantum mechanics the electron B is repelled from A by a force depending on the two electron charges, just as two north magnetic poles repel when brought close together. The strength of this repulsion does not depend on the energy of the incoming electron B. But quantum theory takes quite a different view of A and in particular the "empty space" surrounding it.

We have already encountered the concept of the quantum vacuum in chapter 2. We usually define a vacuum by the absence of anything, but quantum theory renders such a rigid statement operationally quite meaningless. As we have seen, the Heisenberg uncertainty principle says we cannot know the exact position and motion of any particle within an accuracy prescribed by Planck's universal constant. So, we can never say there are *no* particles or quanta of radiation present in a particular region. We cannot have the classical empty "vacuum." In fact, pairs of oppositely charged particles will be continually appearing spontaneously and then annihilating in space, too rapidly to be measured directly because of the veto imposed by the uncertainty principle. These virtual particle pairs characterize the quantum vacuum as a sea of ceaseless activity. Lest one despair of this scenario as an unnecessarily esoteric picture of something that common sense dictates ought to be very simple, it is possible to verify the existence of these virtual pairs by experiment (see page 65).

Creation

Now let us return to the tale of our two electrons. The vacuum surrounding A will be, in reality, not completely empty space but a sea of virtual pairs of other electrons and antielectrons called positrons—none directly observable. The pairs will also be accompanied by photons but since these do not carry electric charges they do not exert forces on the electrons and positrons and we can forget about them. But now, because opposite electric charges (like north and south magnetic poles) attract, the virtual electrons and positrons will become distributed in a special way. The positively charged virtual positrons will tend to be drawn closer to the central negative electron at A than the negatively charged electrons which will be repelled from it. This migration and segregation of charges is called *vacuum polarization* (see Figures 3.1a and 3.1b).

Electron meets electron

When the electron B approaches A it now feels, not the full or "bare" electric charge at A but a slightly reduced version of it that is a little weaker because of the shielding by the virtual charges of opposite polarity. The strength with which B is repelled depends on how fast it is moving. If B approaches A with very high velocity, it will penetrate the screening cloud and get very close to the bare charge at A before being repelled. On the other hand, a slowly moving electron B will barely penetrate the screening cloud and, therefore, will feel a much smaller repulsive force because some of the central charge at A will be effectively cancelled by the oppositely charged virtual particles. The effective

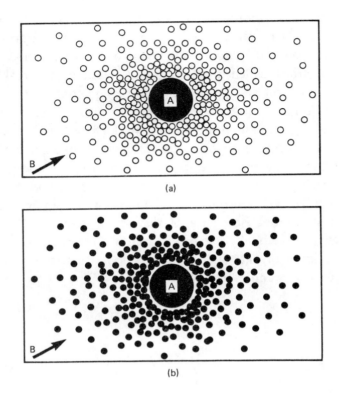

(a)

(b)

Figure 3.1 *Vacuum Polarization.* (a) an electron charge at A is surrounded by a cloud of virtual charges of the opposite sign, and (b) a colored quark at A is surrounded by virtual quarks and gluons with predominantly the same color as the quark at A.

strength of the interaction depends on energy. The electric force seems stronger to incoming particles with high velocity, or equivalently, to those whose temperature or energy is large. These energetic particles will penetrate close enough to A to sample the full unshielded force strength exerted at very small separations between the charges (Figure 3.1a).

Creation

Consider a more everyday example. Imagine we have two solid billiard balls that we surround by thick layers of woollen padding. If these fluffy balls are projected towards each other, the strength of their interaction and the extent of rebound is quite different at low and high collision speeds. If the balls are projected together slowly, their hard cores do not touch and subsequently there is only a comparatively weak rebound through the interaction of their woolly exteriors. If they meet at high speed, the hard cores will penetrate the woollen shields and a strong rebound will occur.

This argument shows why the effective strength of the electromagnetic force depends upon the energy at which it is measured. As the energy of encounter increases, so the strength of interaction slowly increases. Over the range of energies experienced on earth, the change is tiny. For example, as energy changes from the equivalent of 1 up to 10 proton rest masses, the effective electric charge on the electron increases by factor 1.007.

Color

Similar considerations can be brought to bear upon the weak and the strong force strengths. Here, however, there is a dramatic difference. The completely opposite trend ensues. As the energy of interaction increases the strength of the interaction *decreases*. To see why this happens, think about the strong interaction between two identical subnuclear particles, or quarks, A and B. All protons and neutrons contain three smaller particles called quarks. (We will have more to say about them on p. 83). Unlike electrons, quarks have an attribute called color, that determines the magnitude of the strong interaction between them. (This "color" is simply a label for a special quality of the quark. It comes in three varieties and has nothing to do with hues, which are determined by the wavelength of absorbed light.) The strong force acts only on colored particles and is sometimes called the color force. We might have thought that exactly the same thing would happen to our two quarks as with the interacting electrons—a sea of virtual quark-antiquark pairs appears and the quarks distribute themselves to shield the color charge on A from that on the incoming quark B. While this is true, it is by no means the whole story.

The electromagnetic interaction between the two electrons was transmitted by photons that do not carry an electric charge. When photons appear along with the virtual electron-positron pairs, they do not alter the overall electric charge distribution. The strong interaction between quarks, however, is different. The mediating role of the photon is played by a particle called the *gluon*. The gluon mediates the strong interaction, but it differs from the photon in one vital respect. Like the quarks, the gluon possesses the color charge, and so when the virtual gluons appear along with the virtual quark-antiquark pairs, they also affect the color charge distribution around A. (The photons induce no such effect around the electron because they have no electric charge). The effect of the gluon color field is to surround A with a cloud of color charge of the same type as A. This smears out the influence of the color charge at A over a much wider region. Since a smeared-out charge scatters incoming particles of the same charge more weakly than a compact charge at a point, it is clear that the role of the charged gluons is to weaken the effective strength of the interaction between A and B when they get very close together—exactly the opposite effect to that created by the polarized virtual quark-antiquark pairs (Figure 3.1b).

There is clearly an outright competition between the screening of color charge by quarks and the anti-screening by gluons. Which trend wins? The answer is the anti-screening of the gluons. Experiments show that the interaction between the quarks gets weaker and weaker as they get closer together. Calculations confirm this, unless there are many unknown types of quark that would augment the vacuum polarization. At present, it seems more logical to use the experimental evidence to rule out the possible existence of these extra, unfound members of the quark family.

What we have discovered is very, very unusual. As the energy or temperature increases, the strong nuclear force gets effectively *weaker,* all because the gluons which mediate it carry a color charge. The weak interaction is also mediated by particles that carry a weak charge, the so called W boson discovered recently at CERN in Geneva, and a similar effect occurs there. The weak interaction itself weakens slightly at high energy.

Asymptotic freedom

We begin to see how a unification of the forces of nature might arise and where. Because the effective strengths of the interactions change with energy, by going to an environment of very high energy, we may find them to possess equal strengths (see Figure 3.2). The figure shows the calculated change in the effective strengths. We can see that they should indeed all meet at a particular very high energy. Its value is

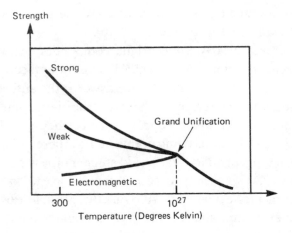

Figure 3.2 *Variation of Forces of Nature with Temperature.* In grand unified theories the strong and weak forces weaken at high temperatures, while the electromagnetic force strengthens. They all have equal strength at temperatures higher than about 10^{27} degrees Kelvin.

about 10^{14} GeV, and corresponds to a temperature near 10^{27} Kelvin. No thermometer known to man can measure such a temperature. These energies are enormously higher than the values obtainable in any terrestrial experiment, only about 10^3 GeV as yet. The only place where such energies can be found in our universe is in the early moments of the "big bang," and it is there that the most dramatic consequences of this grand unification of nature's three basic forces emerge.

Before we go on to investigate further the extraordinary ramifications of this unification, it is worth pausing to see in a little more detail what the weakening of the inter-quark force at high energy means. The property that inter-quark forces possess, that of being weaker at high energy when the separation between the quarks is very small, is called *asymptotic freedom*. This term conveys the notion that if the energies grew infinitely large, the quarks would feel no forces at all. They would behave as completely free particles.

Asymptotic freedom is reminiscent of some elastic forces. If the ends of an elastic band are stretched farther and farther apart, the force pulling them together increases, but as the ends are brought closer together, the forces die away. Think of pairs of quarks joined together by "strings" of gluons. As the separation of the quarks is changed by giving them energy, the string tension alters. When the quarks are pushed close together, there is little tension in the string, and they behave freely—this is asymptotic freedom.

But there is another question to be answered. What if we try and separate the quarks by a large distance? Surely the tension between them should get stronger and stronger. The favored view among particle physicists at present is that the quarks can never be separated by large distances. They, and their color attribute, are "confined" by the tension between them. One of the dramatic consequences of this prediction is that there should not exist any free "colored" quarks or gluons in nature today. They should all reside inside nucleons and other hadronic particles, bound together by gluon strings. Protons and neutrons can be thought of as bags with rubber walls, like balloons. The three quarks within them feel no forces until they try to escape, then they hit the walls of the bag and are pushed back. This seems very foreign to our intuition. We say, why can't we just hit the gluon string joining a quark and an antiquark with an energy high enough to snap it. You can, but the "elastic" energy released on breaking the string is always just enough to make two new quarks joined to the original ones by gluon strings. We are no closer to liberating a quark. This would be like attempting to liberate a single magnetic pole by cutting a bar magnet in half (See Figure 3.3). Cutting the bar only creates two magnets. Most particle physicists suspect that quarks cannot therefore exist in an isolated state. They are always trapped inside particles like protons and neutrons and mesons. For some years, however, a group of physicists at Stanford University, headed by William Fairbank, has claimed to have found isolated elementary particles that, like quarks, possess exactly 1/3 or 2/3 of the charge possessed by the electron. But as yet, no other experiment has been able to confirm these results. It is possible that other fractionally charged particles exist in nature, as yet unsuspected by theoretical physicists and Fairbank has found these. But, if he has really found free individual quarks, some rethinking of elementary particle physics will have to occur.[6]

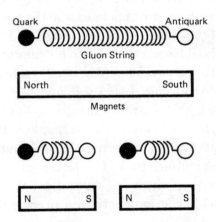

Figure 3.3 *The Analogy Between Quarks and Magnets.* Breaking a gluon string liberates enough energy to produce a new quark-antiquark pair spontaneously. This is analogous to breaking a bar magnet and creating a new pair of magnetic poles.

Creation

Grand unification

What has all this to do with cosmology? We have argued thus far that the behavior we see displayed by the strong, electromagnetic, and weak interactions in our laboratories today is not one we would have anticipated in our history of the early stages of the universe. The properties of these forces depend strongly on the temperature of the environment in which they are acting. By the standards of the "big bang," we live in a very cool world indeed, and so the behavior of matter would be very different now from what it was during those early moments of the universe.

This presents us with a very exciting prospect. Although our terrestrial particle accelerators cannot attain the huge temperatures and energies necessary to probe the new partially explored events surrounding the epoch of grand unification, by working out the cosmological consequences of such a theory, we can test it against astronomical evidence. We might then witness not only a unification of the forces of nature but a unification of the sciences of nature.

The property of asymptotic freedom is a godsend to cosmologists. It means that at higher and higher energies interactions between elementary particles get weaker and weaker, and this brings a study of the extreme environment of the "big bang" within our current capabilities. If interactions were to grow ever stronger at high temperatures, we would encounter fantastic complexity on approaching the earliest cosmological moments. Everything would degenerate into one continuous elementary particle interaction. Encouraged that we have found a route into the problem, let us see how close to the beginning of the universe it leads us, and what awaits us when we get there.

The Left Hand of Creation

The picture we imagine of the early evolution of the universe is colored by our new discoveries in particle physics. Our study of the universe begins at the earliest intelligible instant, the Planck moment when the temperature is 10^{32} Kelvin, and there is complete symmetry existing between the strong, electric and weak forces—the strengths of their interactions are all equal. As the universe cools, the different interactions distill off and become distinct in strength and range. At first, after about 10^{-35} second, when the temperature falls to 10^{27} Kelvin, the strong force becomes distinct. Then much later, after about 10^{-11} second, the electromagnetic and weak forces emerge to be quite different forces of Nature as we now see them (see Figure 3.4).

So far we have seen how physics in the early universe is quantitatively different. That is, the strengths of forces between particular forces differ from those we measure today.

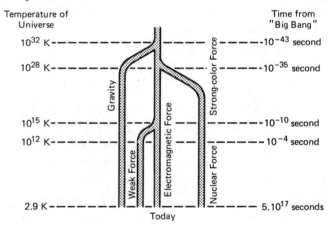

Figure 3.4 *Symmetry Breaking.* The evolution of the breaking of symmetry between the different forces of nature. As the universe ages and cools the strengths of the forces become dissimilar.

SOURCE: H. Sato, "The Early Universe and Clustering of the Relic Neutrinos," in A. Ramaty and F. Jones, eds., *Tenth Texas Symposium on Relativistic Astrophysics,* vol. 375 (New York: Annals of the New York Academy of Sciences, 1981), 44.

Creation

The most exciting differences, however, are actually the qualitative ones—the evolution in the types of existing particles and their interactions. We recall that, besides the differing strengths of nature's forces, the other superficial barrier to unification of the forces is the different type of elementary particles that they each influence. In the low energy world of our physics laboratories, the strong nuclear force controls the behavior of those elementary particles like quarks and gluons, which possess the attribute of color. The weak and electromagnetic interactions affect the behavior of a distinct class of particles, called leptons, meaning "light ones," including the electron, and all the different neutrinos. The neutrino received its name from a comment by the Italian physicist Enrico Fermi. It was "little neutron": neutral, like the neutron, but much lighter, indeed perhaps without any mass at all. Its properties will be further described below.

Leptons do not feel the nuclear force because they do not carry the property of color. Color, like electric charge, is a conserved quantity, and so the total value of the color possessed by a group of particles is a conserved quantity. The total value of the color at the end of the interaction must be the same as it was at the beginning, although the number of colored particles can, of course, change. Because of this, there appears to be no way of transforming colored quarks into colorless leptons and vice versa at low energy. If a true unification of these different interactions of nature is to occur, then we require an intermediary that possesses color.

X particles

This colored mediator of quark-lepton transformation exists in a unified theory and this mediator is called the X boson. The X boson is a particle like the photon, but it is extremely massive and requires huge amounts of energy to produce. The reason we see the strong and electric forces acting on different classes of particles in the laboratory is because at the very low temperature and energy of the experiments, it is very improbable that any X particles are created. Only when they are abundant will we tend to see the unification of quarks and leptons which they mediate. Because Xs carry both color and electric charge, it is possible for the bookkeeping of color and electric charge to balance at the end of a transformation of quarks to leptons. The energy necessary to make an X boson is about 10^{15} times larger than that necessary to create a proton and is also, as we would expect, about equal to the energy at which the effective strengths of the strong, weak, and electric forces become equal. When the universe was hotter than this, about 10^{27} Kelvin, the abundance of X bosons was profuse, and they mediated continuous transmutations between quarks and leptons. The strength and the object of all the different interactions are identical in this exotic era when the universe was less than 10^{-35} second old.

As the universe expanded and cooled, so energies fell. The X bosons gradually disappeared and the strengths of the different interactions decayed at differing rates. The universe evolved from a symmetric state where the various forces of nature were identical, into one of asymmetry where different interactions act exclusively on separate classes of particles.

Creation

Before we see how dramatic a role the X particles may have played in the creation of the universe, we must enlarge on something rather unusual—antimatter.

Antimatter

In 1927, the English physicist Paul Dirac, realized that for ordinary matter to be stable, there must exist a sort of mirror image of it. This complementary form of matter, called *antimatter,* possesses equal and opposite values of the various labels associated with elementary particles. For the electron with its unit negative electric charge, there must exist a positron of equal mass but unit positive electric charge. Likewise, for the proton, there is a negatively charged antiproton. Five years after Dirac's dramatic prediction, a positron was detected in the laboratory by Carl Anderson and finally, in 1955, Owen Chamberlain and Emilio Segre at Berkeley found an antiproton. Today, the appearance of antiparticles in high energy physics experiments is commonplace. But despite the ease with which antimatter can be produced artificially, one great cosmological mystery remains: why do we never see antimatter in space? The solar system contains no anti-planets. Nor are the neighboring stars and galaxies or the diffuse material between them made of antimatter. If they were, the flood of gamma rays that result from the catastrophic annihilation between matter and antimatter at the boundary where their two domains meet would be clearly visible. Prolific gamma radiation is not found; the flux of gamma rays in the universe is extremely weak. This testifies to the scarcity of antimatter in space. Why should such an overt cosmic favoritism for matter over antimatter exist in the universe?

Traditionally, there have been just two answers to this awkward question. To appreciate them, we must first understand why the question is awkward. There has always appeared to exist a rather precise and elegant law of nature associated with the relative imbalance between matter and antimatter in a closed physical system. Consider a collection of baryons, particles like protons and neutrons, and their antiparticles, and assign the number $+1$ to each of the particles and -1 to each of the antiparticles. Now total up all the numbers for the system: the result is called the baryon number. (Leptons do not possess a baryon number). The particles can interact, transmute, or decay, provided we assign values $+1$ or -1 to the members of the system at any moment and total them up. We would find the final total, or baryon number, to be always the same. This is a simple example of what physicists call a conservation law (other examples of conserved quantities are the total energy and electric charge of the particles in the system). Now, if baryon number really is a conserved quantity, then the very large baryon number in the universe (we observe at least 10^4 particles with $+1$ for every antiparticle with -1 in space) must have been built into the "big bang" at the beginning of time. No physical process could have altered its total value subsequently; there has always been an overwhelming preponderance of matter over antimatter, and none of our theories can explain it. It is a created property of the universe.

Creation

Cosmic favoritism

To the physicist, this is an unattractive conclusion. Not only do we have to parcel up this remarkable fact about the universe and mark it unalterable and unexplainable, but we have to face up to the awkwardness of its value. We can look at this value in another way. The fact that today we observe about two billion photons in the universe for every proton, on the average, means the "big bang" had about a billion and one protons emerging from it for every billion antiprotons. The antiprotons annihilated with their billion anti-partners to create about two billion photons for every leftover proton. If we are going to make special assumptions about the beginning of the universe, it would be more appealing to assume complete equality between the number of particles of matter and antimatter rather than 1,000,000,001 particles for every 1,000,000,000 antiparticles.

Many astrophysicists once favored the idea that the natural baryon number for the universe is zero—that is, complete symmetry between matter and antimatter. Because our surroundings are almost exclusively matter rather than antimatter, we must appeal to an extremely nonuniform distribution of matter and antimatter to explain why we do not see equal amounts of matter and antimatter. Some parts of the universe are all matter, others perhaps all antimatter. We, evidently live in a matter-only section. Unfortunately, this seemingly attractive scheme does not work. The universe expands, and so in the past, matter would have been more densely distributed and the matter and antimatter domains in closer contact. The result: catastrophic annihilation. A universe beginning with equal quantities of matter and antimatter would have resulted in annihilation occurring so efficiently during its early history that we would see 10^{18} photons for every proton or antiproton today. This prediction conflicts with the observed photon background by a factor of a billion. It appears we live in a world with a "hidden" matter-antimatter asymmetry. The universe exhibits a favoritism for matter over antimatter.

One of the most dramatic consequences of the grand unification theory of elementary particles is that it provides a resolution both of this conundrum of matter and antimatter imbalance in the universe and offers an explanation for the magic number of one billion or so photons for every proton observed in intergalactic space today. Before we see how this is possible, we should mention another dramatic role played by the X boson. This role not only enables us to verify its existence, but allows us to test the possibility of baryon number nonconservation as well. [7]

Proton decay

The protons of which all atoms and molecules partially consist are not really elementary particles at all. When protons collide together, they behave as though they possess internal constituents. These subcomponents are the quarks we have already met. Each proton contains three quarks. They carry baryon numbers of 1/3, 1/3, and 1/3, to yield a total of +1 for the proton. Recall, however, that the X boson is able to mediate transmutations of quarks into leptons—particles which have zero baryon number—and hence such transmutations violate the baryon number. It is now possible for two of the quarks within the proton to decay into a positive electron and an antiquark. The latter pairs with the third quark in the proton to form an observable particle called the pi meson or pion. This process is extremely rare in the present universe because the ambient temperature is too low for X bosons to be very abundant. Occasionally, a large fluctuation of energy occurs in the micro-world which is large enough to create an X-boson and stimulate a proton to decay. On the average, one has to wait about 10^{31} years for a single proton to decay by this means. This seems at first an absurdly long time and of no interest to experimentalists— after all, the age of the universe is a mere 10^{10} years. But there are so many protons in large quantities of matter that it is quite practical to detect these rare decays. One thousand tons of water contains enough protons to make a couple of decays a month likely. To shield out unwanted cosmic ray particles from the detectors, a number of experiments have been set up in deep mines below the surface of the earth. Two different experiments have tentatively shown some events with the characteristic pattern of proton decay; a third experiment has not. If the decays have been found, then all things, even diamonds, are not forever. All nuclear matter is unstable. Atoms will eventually all decay away. The immensity of the proton lifetime is equivalent to about one proton in your body decaying during a typical human lifetime. Even if the lifetime were shortened to about 10^{16} years—still a million times the age of the universe—the biological consequences would be catastrophic. The decays would deliver a lethal dose of radioactivity in less than a year. [8]

Although the effects of the X particle are barely detectable in our present rarefied universe, they were far more rapid and dramatic during the first instants of the "big bang." There, the extremely high temperature ensured they would exist in profusion. Because X particles mediate particle interactions, which change the matter-antimatter (baryon number) balance of a system, it is possible that they can explain precisely why the universe is not matter-antimatter symmetric today.

Creation of matter

When the universe was about 10^{-35} second old, X particles along with their antiparticles would have been profusely produced in equal numbers, so high was the temperature of the universe. A few moments later, the X and anti-X, (or $\overline{X}$), particles began to decay into quarks and leptons by the same process that causes protons to decay. However, the decay rates of the X and the $\overline{X}$ are different, and the result is that *equal* numbers of X and $\overline{X}$ particles can decay into *unequal* numbers of quarks and antiquarks. Later, as the universe cooled, these quarks matched up in triplets to form protons and antiprotons. The excess of quarks over antiquarks eventually manifested itself as an excess of protons over antiprotons. Annihilation between protons and antiprotons followed, and a particular ratio of protons to photons emerged.

The final ratio of protons to photons depends on just three quantities. First, the fraction of the universe that is in the form of the X and $\overline{X}$ particles to start off with. This is found to be about one percent. The second quantity is the difference in the X and $\overline{X}$ decay rates, which is very difficult to calculate. (The decay rates may one day be measured in the laboratory because the same rate difference may be responsible for creating the detailed charge distribution inside the neutron. Although the neutron has no overall electric charge, hence its name, it is expected to possess a tiny difference in charge between any two of its hemispheres.) Finally, there is an "efficiency factor" that measures the X decay rate relative to the cosmological expansion rate. The product of all these factors predicts the photon to proton ratio could lie anywhere between about 10^{+4} and 10^{+13}, and so could indeed explain why we see 10^{+9} photons per proton in the universe today. In the future, the uncertainty in this prediction should be honed down by better calculation and experiment.

The preponderance of matter over antimatter in the present universe and the relative balance of matter and radiation we observe in space appear to be artifacts of events occurring in the first 10^{-35} second of the universe's life. The asymmetric decay of the X particles is complete just 10^{-30} second after the expansion of the universe commences. At that point the X particles already have largely determined the fate of the universe.

The imbalance between the number of quarks and their antiparticles is reflected in the small number of particles of matter that survived the last annihilations of matter and antimatter and fashioned the cosmic structures we see today. The exotic processes that play a role in establishing the large-scale structure of the universe are predicted to have several tiny effects elsewhere that are just within the range of our most sensitive proton decay detectors. Until very recently, cosmologists never dreamed of having such good fortune.

Quarks are confined

The matter-antimatter imbalance of the universe is established 10^{-35} second after the beginning and thereafter the universe of quarks and antiquarks continues to cool as the expansion further rarefies the cosmic medium. We think that no real drama will be played out on the cosmic stage until the temperature has cooled to about 10^{15} Kelvin. By then, about 10^{-11} second will have elapsed, and a new act will begin: The temperature will be so low that W and Z particles, the lighter analogues of the X particle that mediate the weak and electromagnetic interactions, will no longer be made in profusion. These particles are responsible for maintaining the symmetry between the weak and electromagnetic interactions, just as the Xs did for the strong and electroweak interactions, but are unable to create a matter-antimatter asymmetry. When they disappear, then all electromagnetic and weak interactions will begin to look different. The reason we think radioactivity and electricity are different forces is again because we live in a cool universe.

Soon after the W particles disappear, an extraordinary transition occurs in the cosmic medium. This change transforms it from the rich and strange world of elementary particles into the more familiar world of protons and neutrons, the constituents of atomic nuclei. When the temperature falls to about 500 MeV, after a millisecond of cosmic expansion, the cosmic soup of quarks and antiquarks and photons suddenly thickens. The temperature is low enough for the process of confinement to occur. Remember quarks interact more and more weakly when brought close together, but when they are not very energetic, and so widely separated, an increasingly strong restoring force pulls them together. After a millisecond of expansion, the universe is cool and rarefied enough for quarks to separate sufficiently to notice these confining forces. Instead of just separating continuously along with the expanding, rarefying universe, the attraction binds them tightly together into pairs and triplets.

Creation

Then combinations of protons and neutrons that compose everything around us and within us materialize. At that moment, the tiny imbalance of quarks over antiquarks we have inherited from the earlier epoch of grand unification is reflected in a surfeit of protons and neutrons over antiprotons and antineutrons. The antiparticles quickly annihilate their partners leaving about one proton and neutron surviving for every billion photons of annihilation radiation.

The first second

The universe is now a millisecond old. It is full of protons and neutrons amid a sea of neutrinos, electrons, positrons, and light. It is becoming a world of more familiar particles, but all in unfamiliar states. All the complex nuclei we now encounter are built of protons and neutrons. Only hydrogen is just a single proton. If we are going to explain why the universe contains particular abundances of all the different nuclei, we clearly have to explain what fixes the relative abundance of neutrons and protons at this stage of the universe's expansion.

Until 1951, this was an unsolved problem. It appeared that there was no special ratio of neutrons relative to protons in the "big bang" picture, and, therefore, a special initial ratio would have to be assumed. Then Chushiro Hayashi, a Japanese astrophysicist, made a key discovery. He found that from the moment quarks condense into nucleons until the universe is about a second old, the neutrinos start to play a leading role.

The neutrino was once described as the nearest thing to

nothing yet conceived of by physicists. This description was provoked by the fact that the neutrino does not participate in the strong and electromagnetic interactions. It feels only the influence of gravity and the weak interaction. So weak are these neutrino interactions that many billions are passing through the reader's head at this very moment, with no ill effects. In order for us to catch even the tiniest fraction of the neutrino hordes issuing from the center of the sun, we must bury detectors deep in the earth so that they pass through sufficient material to make it possible that a few will be stopped in the detection area and at the same time insure that the detector will be shielded from the myriad of other cosmic particles.

But in the first second of the universe's history, temperatures and densities were so high that even neutrinos would undergo frequent interactions with other particles. Hayashi realized that these interactions—similar to radioactive beta decays—would solve the neutron-proton puzzle for us. When less than a second old, the temperature of the universe is higher than 10^{10} Kelvin and the neutrinos mediate transmutations of neutrons into protons and vice versa. These reactions happen so rapidly compared to the rate at which the whole universe is expanding that the number of protons and neutrons is kept precisely equal. Any slight excess of one particle type just leads to an increase in the production of the other, and equality is reestablished. But, the neutron (1.67492×10^{-24}g) is minutely heavier than the proton (1.67266×10^{-24}g) and so requires a little more energy to produce. So, gradually, as the universe cools towards 10^{10} Kelvin, the protons become very slightly more abundant because they are easier to make. Finally, when the temperature reaches 10^{10} Kelvin, the density is too low for the neutrino interactions to keep pace with the expansion of the

Creation

universe and the transmutations cease. The neutron-proton balance is fixed, and the neutrinos become the virtually free particles they are in space today. This sequence of events precisely determines the relative availability of neutrons and protons in the early universe.

Remarkably, we see that the answer does not depend upon the unknown events at the "beginning" of the universe but only on conditions at temperatures of 10^{10} Kelvin—quite moderate by the standards of high energy physics. In this regime we have an excellent, well-tested understanding of physics. The average density of matter in the universe at these moments is less than that of water, in fact.

Light element synthesis

If nothing further had intervened, cosmic evolution would have soon become very mundane. After about 926 seconds, a free neutron undergoes a radioactive decay into a proton, an electron, and an anti-neutrino. The material universe would contain only protons—pure hydrogen nuclei (it is still, at this early time, far too hot for atoms to form). Before these radioactive decays can occur to remove the neutrons, however, they feel the effects of the strong nuclear force. The protons and neutrons combine very rapidly to form deuterium and then helium. During a small niche of cosmic history, between about 10 and 500 seconds, the entire universe behaves as a giant nuclear fusion reactor burning hydrogen to helium. Previously, no helium nuclei could exist. The temperature was so high that the sea of radiation smashed them back into the pieces from which they were built. Subsequently, the density is too low for the strong interactions to occur; protons and neutrons cannot get close enough to overcome the electromagnetic repulsion of similarly charged nuclei and bind together.

We can predict the percentage of the universe that should have emerged from this nuclear era in the form of helium. About 25 percent of the mass of the universe ends up in the form of helium, about 75 percent remains as hydrogen, with a minute 10^{-3} percent in the form of deuterium, and less than one millionth of a percent as lithium. All have been found in space with the predicted abundances, to within the observational uncertainties.[9]

When we measure the abundance of helium in the universe, no matter where we look, whether it is in the oldest stars in our galaxy or in remote galaxies, we find this prediction confirmed. There exists a universal abundance of helium at about the 22 to 25 percent level. Even in sites where the other heavier and rarer nuclei, like carbon, nitrogen, and oxygen, have reduced abundances compared to the average, helium still displays its universal value. What these observations are suggesting is that, whereas the heavier elements were made more recently in local objects like stars and so reflect the different conditions from place to place, and star to star, the helium has a universal origin and its abundance will not reflect the nature of the local environment.

Over the last fifteen years, as better and better observations of helium in interstellar space have been possible, the "big bang" predictions have been borne out to greater and greater accuracy. Deuterium has even been detected in interstellar clouds. A remarkable property of deuterium is that stars only destroy it; they do not create it. The sole source of deuterium seems to be the "big bang." Therefore, its detection, along with that of helium, provides impressive experimental evidence that our model of the early universe is substantially correct back to just one second after the "big bang." Models of the early universe based on general relativity and the relevant aspects of elementary particle physics can potentially explain the existence of matter, and the absence of cosmic antimatter. They can explain not only the relative number of photons to particles of matter, which measures the rate of entropy production over the entire history of the universe, but also the material and chemical composition of the observed universe.

At CERN, the LEP experiment has shown that there are precisely three species of neutrinos: a larger number would have added energy density to the early universe and thereby slightly speeded up the expansion. This would have led to the overproduction of helium by a small but significant amount. The primordial nucleosynthesis of the light element abundances predicts that the number density of baryons in the Universe must lie between 2 and 10 times the number density known to be in the form of luminous stars. The existence of a substantial amount of baryonic dark matter, perhaps enough to account for the dark matter around galaxies, is thus a prediction, as yet unconfirmed, of the Big Bang theory. This quantity of dark baryons is insufficient to give the density level predicted if inflation occurred in the very early universe. Thus, inflation predicts the existence of additional, non-baryonic dark matter.

4

Evolution

STRUCTURE is the essence of life. A featureless universe offers a vision of a hostile and arid environment, devoid of the oases that allow life to be nurtured and flourish. Our solar system began as a whirling cloud of gas that cooled and condensed into the planetary bodies. Supply a little initial structure, and its subsequent development is inevitable under the relentless pull of gravity. Whether the complex evolutionary chain successfully culminates in man is not our immediate concern. Rather, we gaze in awe at the splendor of a starry night sky, and ask: Where did stars such as our sun come from? The jigsaw puzzle of cosmic structure points to our galaxy as the source. Studies have revealed the age of many of the stars around us. The oldest members of our stellar society have been identified. They bear witness that our Milky Way galaxy came into existence about fourteen billion years ago. (This age estimate is uncertain by some five billion years.) Such an answer is unsatisfactory, of course, because it only defers our quest for the origin of things by raising the still more profound issue of the origin of the galaxies themselves. Observation provides us, as always, with the vital clues as to the origin of the large-scale structure of the universe in the remotest depths of time. The knowledge gained in the attempt to understand the origin of galaxies, of clusters of galaxies, and of the great superclusters of galaxies leads us ever closer to the beginning of the universe itself.

The richness of the structure in the universe readily becomes apparent even to the casual observer. View a packed football stadium from afar, and the crowds appear remarkably uniform. Only on closer inspection does one notice a great diversity. Some sections of the terraces are tightly packed; the better seats contain a lower density of people; behind one goal, the supporters predominantly wear blue; behind the other, the prevailing color is red.

And so it is with the galaxies. Peering through a small telescope from our earthly grandstand, the cosmos seems remarkably homogeneous. Looking through a telescope of greater aperture and resolving power, we begin to notice irregularities in the cosmic landscape. Finally with a very large telescope, we see far beyond our Milky Way. Galaxies are sprinkled everywhere. We now begin to gain a perspective on the true structure of the universe.

Structure

In recent years, Hubble's red shift–distance relation has been used to calculate distances to thousands of galaxies.

This has led to a major advance in our mapping of the cosmos. Galaxies are projected on the surface of the celestial sphere to form a two-dimensional picture. By using systematic distance measurements, we can develop a true three-dimensional picture of the distribution of the galaxies in space. It reveals unexpected features. Galaxies are mostly concentrated in filamentary and sheet-like structures, thin compared to their typical separations of 100 to 400 million light years. More striking still, these surfaces enclose large regions of space that are almost completely devoid of any luminous galaxies. The layers where galaxies are found are not uniformly populated. There are great clusters and super-clusters of galaxies, seen as localized peaks in the galaxy distribution and often occurring where the filaments intersect. The human eye can easily recognize these intricate patterns even though a quantitative analysis is not readily forthcoming (see Figure 4.1).

Gravity is undoubtedly responsible for the formation of galaxies and their unusual distribution in space. Our first guess about the effect of gravity is that it concentrates matter in roughly spherical lumps of different sizes which are completely at random. This simple picture does not resemble the real universe, however. Do we need to add new forces or very artificial initial conditions to explain the actual observations? Not necessarily. Gravitation theory alone is capable of providing a natural explanation for both the characteristic size and pattern of the galaxy distribution. No special initially imprinted scales, patterns, or new forces seem to be required.

Evolution

Primordial seeds

There is a unique rate at which fluctuations in density can grow under the effects of gravity. Actually, it is the extra pull of gravity associated with the slight excess of particles relative to the smooth ambient background density that causes any little fluctuations to grow more and more pronounced with time. Unfortunately, in an expanding universe, the density of matter is continuously decreasing, on the average, at precisely the same rate as this growth. For only one time-scale of change naturally enters into the behavior of a system where gravitational forces dominate, and this time-scale is completely known once the matter density is specified. Not only is the mean density of the universe decreasing with time, but so also is the slight excess associated with fluctuations. Consequently, the fluctuation has a lot of difficulty in attempting to become more and more pronounced. It is rather like trying to win a race on an ever-lengthening track. The fluctuations can still grow, but relatively slowly, since the time-scale over which they can amplify is precisely the time-scale over which the background density is decreasing. The upshot of this is that for large fluctuations to have been present in the past, at the epoch of galaxy formation, there must have been some small fluctuations present initially near the beginning of the universe.

Now this may not seem much of a requirement. Yet understanding the origin of these tiny fluctuations poses one of the greatest challenges to modern cosmology. Any primordial fluctuations that do develop could not have been too dramatic, for had the very early universe been grossly inhomogeneous, it could not have evolved to the observed state of

Figure 4.1 *The Distribution of Galaxies.* A computer simulation of a random distribution of galaxies *(left)* and the actual observed universe from a map of the galaxy distribution *(right).*

regularity today. Evidently, the balance between chaos and uniformity was rather delicate.

The near-uniformity of the universe raises another, perhaps even greater paradox. There is a limit to the effective scale of the universe over which fluctuations from perfect uniformity can be created or destroyed at any epoch in its history. This is simply the distance that light can have traversed since the beginning of the expansion. This distance, which of course increases inexorably with time, represents the maximum size of regions that could have been in causal contact. Such contact is necessary if we wish, for example, to appeal to physical (rather than metaphysical) processes, either to create spontaneously fluctuations or to smooth out any overly large pre-existing inhomogeneities. We call this causal limit the horizon size of the universe.

At the threshold of classical cosmology, which we take to be the Planck instant, the horizon size encompassed no more than the mass of a grain of sand! But, the seed fluctuations which gave rise to the large-scale structure we see today must have been present on scales at least as large as those corresponding to the masses of highly luminous stars. Even this stipulation presumes some sort of enlargement of the irregularity, such as might occur when massive stars explode and sweep up large shells of ambient matter, in order to amplify structure all the way up to galactic scales, and indeed supergalactic scales. More probably, the seed fluctuations were themselves of galactic scale, encompassing a hundred billion suns.

Such density fluctuations, if arising spontaneously, like the erratic condensation of lumps in a saucepan of unstirred custard, would have to form at a rather late epoch, when the universe is more than a year old. The relevant physics is well understood, and does not allow any spontaneous generation of fluctuations over the dimensions of interest for galaxy formation. Likewise, the observed isotropy of the Hubble expansion is hard to understand in a universe where regions that are identical now were, until relatively recently, disjoint and out of contact for all practical purposes. None of this is very satisfying from an aesthetic point of view. One recourse is to appeal to a very special set of initial conditions for our universe, but this largely begs the question. It amounts to a modern version of Gosse's trick of having things only take on the appearance of being old while in reality being young.

Inflation

An exciting development in particle physics has emerged to rescue cosmology from this teleological quandary. During its brief passage through a very high energy state, we saw the matter content of the universe undergo a transition between two distinct phases. During the first phase, the energies of particles were so great that a grand unification of the different interactions of nature arose. In the following, less energetic phase, the strong nuclear force decoupled from the weak and electromagnetic forces, and the symmetry of the interactions between leptons and quarks was broken. This phase transition can be compared with the melting of ice. Hidden energy (or latent heat) maintaining the structure of ice molecules is released as it takes up a less rigid, liquid, form. Similarly, the soup of elementary particles that compose the early universe can exist in different phases, and a gradual change from one state to another can release considerable energy. The heat released represents the amount of energy required for the transition from the lower to the higher energy state.

The key to evaluating the significance of the phase change is the actual amount of energy release. If the energy release is sufficiently great, it can have drastic consequences for cosmology. For when it occurs, the previously decelerating Friedman universe could suddenly accelerate. The separation between any pair of points now increases at an exponential rate. Horizons rapidly enlarge, and the universe would open up its secrets. But the phase transition is thought to occur smoothly and uniformly, and the exponential expansion lasts only for a short period until the transition is completed. Then the Friedman character of the universe reasserts itself, and expansion resumes at its previous, more leisurely and decelerating pace.

There are some remarkable consequences for this post-inflationary universe. Its passage through an epoch of broadened horizons admits the possibility of understanding the horizon paradox, whereby initially disjoint sectors of the universe now appear uniform. The size of the universe can be understood. It is as large as it is today compared to its size at the Planck instant simply because it has inflated. But most amazing of all perhaps, the inflationary universe may resolve the mystery of the origin of the fluctuations from which galaxies eventually grew. Fluctuations inevitably exist on the microscopic quantum level because of Heisenberg's uncertainty principle. The inflation simply amplifies the extent of these fluctuations, just as the inflation of a balloon amplifies the area spanned by blemishes in the rubber fabric. No particular size of fluctuation is favored. The inflating horizon acts democratically, leaving behind fluctuations of similar strength over progressively increasing distances. The amount of gravitational energy is the same on each scale up to a maximum size, which corresponds to the effective horizon just when the inflationary phase is ending. This scale can greatly exceed the size of the observable universe today. Clearly, everything hinges on how much inflation there is, how long it lasts, and how suddenly it ends. These are some of the things cosmologists are trying to determine. We can find a model that has everything just right in theory, but we need to prove that this model necessarily describes reality, because we have also been able to find models with both too much, and too little, inflation. These are the models we would like to exclude. Only if a theory can predict a fluctuation level of between one-tenth and one-thousandth of a percent will it be possible to say that galaxy formation is inevitable. The reason cosmologists are so keen to confirm the general idea of the inflationary universe is not just because it can resolve the problems discussed here. In the next chapter we shall see that it has a few more tricks still up its sleeve. [10]

Reheating

The universe is not completely empty. Excessive inflation could be disastrous. It would only solve one problem at the expense of creating an even greater one. Fortunately, the enormous kinetic energy associated with inflation is eventually converted into heat. The universe gradually reheats as the transition occurs and must do so sufficiently early for grand unification to again play a role in coupling strong and weak forces together and for X particles to reform and decay. The continuing expansion cools the matter; the exotic particle interactions cease, but a residue of baryons is left behind. These baryons are finally those we now identify as characterizing the matter around us.

The ratio of the number density of baryons to that of photons is approximately independent of the epoch and provides a convenient measure of the matter content of the Universe. This ratio was imprinted when particle energies finally cooled below those at which grand unification could occur. Subsequently, one could only conceive of diluting this ratio if the entropy of the universe were to increase as the result of photon creation. Processes such as dissipation, further phase transitions, or the decay of particles that are hotter than their surroundings could lead to substantial entropy production. In the absence of such effects, we say that the expansion is entropy-conserving or adiabatic.

Fluctuations in the curvature of space

Now the grand unification theories guarantee that if the usual "big bang" theory were applicable, there could not be any significant spatial variations in the ratio of baryon to photon number densities. The only allowable types of energy density fluctuations are those which preserve this ratio. These are referred to as adiabatic fluctuations and involve the total energy density, including radiation and neutrinos as well as matter. Energy density fluctuations correspond to variations in the geometrical curvature of space-time. According to Einstein's theory, density inhomogeneities are a source of gravity and are equivalent to deviations from Euclidean geometry. The "big bang" cosmological model itself has a geometry that is not necessarily Euclidean. Any individual fluctuation behaves rather like a slightly perturbed "big bang" model at these early epochs, and so the curvature structure of the early universe can be thought of as possessing small wrinkles in its geometry. In other words, adiabatic or energy density fluctuations and curvature fluctuations are equivalent concepts.

Inflation can produce these geometrical wrinkles over a wide range of possible scales. Whether or not inflation really is their source it would seem entirely fortuitous that our theory should single out the scale, say, of a galaxy. As the expansion proceeds, larger and larger amounts of mass come within any observer's horizon, which provides a natural scale for measuring cosmological curvature fluctuations. Once a galaxy mass is first contained within a horizon, we can describe it adequately by ordinary Newtonian concepts of gravity. The directly observable density fluctuations are all we need subsequently consider.

Instability

Isaac Newton qualitatively, and then James Jeans quantitatively, realized that perturbations in a self-gravitating gas are unstable. Density fluctuations grow larger and larger as time passes. This has two competing effects. Gravity attracts the particles towards the highest densities, while the chaotic motions and pressure gradients try to prevent this. In fact, pressure is the manifestation of chaotic motion. On a large scale, gravity always wins, matter collapsing in some regions and becoming rarefied elsewhere. The phenomenon occurs for all kinds of material, even radiation, since all are subject to gravity. The tendency towards instability remains for the behavior of even those density perturbations that occur in the expanding universe. Pressure effects are important over small scales, where the perturbations can be visualized as slight density enhancements that propagate like sound waves passing through air. The excess density is accompanied by a corresponding increase in the pressure, which in turn, induces a further disturbance in adjacent volumes of the fluid. Any given region of enhanced density moves at the local speed of sound. An observer at any point would view an oscillation in the density as the disturbance propagated past him.

On very large scales, and always outside the horizon where not even motion at light speed has any effect, gravity has always provided the dominant force. There has not been sufficient time for pressure forces to be exerted since this requires propagation of a stress at the sound speed. However, it would require a velocity exceeding that of light to drive a coherent motion on a scale larger than the horizon, and the speed of sound cannot exceed the velocity of light. The consequence of this is that on sufficiently large scales, matter and radiation distributions still remember their past. There has not been time enough to erase the lingering memory of the initial curvature fluctuations.

Radiation drag

Until the temperature drops to a few thousands of degrees Kelvin, some 300,000 years after the "big bang" singularity, the quanta of radiation remain sufficiently energetic to keep matter ionized. It takes a minimum energy of 13.6 electron volts for a photon to ionize a hydrogen atom, that is, to turn it into a free proton and electron. Only when practically all photons have been reduced in energy by the cosmic expansion to below the ionization threshold do hydrogen atoms subsequently become the predominant form of matter.

During the ionized phase, the scattering of free electrons by the radiation provides a strong source of friction. From the perspective of an electron, it is rather like trying to run through very dense undergrowth. Any motion of protons and electrons is inhibited relative to the radiation, and fluctuation growth by gravitational instability is inhibited. This drag by radiation on matter is compounded by actual erasure of the smaller-scale adiabatic energy density fluctuations. One of the components of such a fluctuation is the compressed radiation. Radiation always has a tendency to diffuse away. As it does so, it effectively smooths out all fluctuations from which there is time for the radiation to leak out. One finds that fluctuations are completely erased below a scale of about 10^{14} solar masses. This corresponds to a distance at the present time of about 50 million light years. Primordial structures that arise directly from the seed fluctuations will only survive over larger dimensions. Now the radiation is scattered predominantly by electrons during the radiation era. Once the electrons recombine into hydrogen atoms, the radiation then moves independently of the matter. There is no longer any resistance to growth of matter fluctuations. Gravitational instability can now proceed with full vigor.

Sheets and voids

The absence of pressure has a dominant effect in determining the structure and shape of the first objects to form. Thermal pressure is always isotropic, and if pressure is comparable to gravity, one expects the formation of objects with near-spherical symmetry. Only if pressure is completely negligible up to the last moments of collapse, can and will highly elongated or flattened structures develop. Furthermore, nearly all of the matter will collapse into the compressed high density regions since there is no pressure to counteract the infall. A simple argument shows how efficient this collapse is likely to be.

There are three perpendicular directions of space relative to which any motion can be referred. The probability that matter will be compressed or rarefied along any one axis is one-half. The fraction of gas which will not be compressed along any of these independent axes, therefore, amounts to about $0.5 \times 0.5 \times 0.5$ or one-eighth. This has immediate implications for the spatial structure predicted for the final state of the universe. At any early stage, when the density was still nearly uniform, let us imagine boundaries around the regions destined eventually to be compressed and form galaxies. Suppose these regions contain about 90 percent of all matter. Initially, they surround smaller bubbles of matter that never collapse. The bubbles are destined to become huge voids. A metamorphosis occurs as the compression develops. The small bubbles, containing only 10 percent of the mass, become rarefied and now occupy 90 percent of the volume, while the density excesses have become flattened into sheets or filaments. Eventually, the compressed regions, which initially enclosed the smaller expanding regions, surround the voids, although they only fill a small fraction of the volume. The mixture of thin walls and filaments of compressed matter resembles a cellular structure dominated by the presence of huge voids.

Pancakes

Early on, soon after the atomic era commenced, when matter was predominantly in atoms of hydrogen, the perturbations in density were small, with fluctuations of less than a percent. The effects of radiation smoothing had strongly suppressed any small-scale structure. As we follow the trajectories of particles perturbed in this way, undisturbed by pressure forces, the particle paths eventually intersect. Just as any ordered motion of spectators at a crowded football game results in regions of intense crowding, so will the particles we are following fall into surfaces of high density.

Consider a small cubical volume of the universe. One can visualize this as being deformed under gravity, with contributions both from the cosmological expansion and from the self-gravity of the local density enhancement. The deformation of the cube can be regarded as a tendency to collapse or expand along each of three perpendicular axes. Clearly, a spherically symmetrical collapse will be a very special case. Both the sense and the magnitudes of the three independent deformations along each of the axes, into which any general collapse can be decomposed, have to be synchronized in order to yield a spherical collapse. Imagine trying to coordinate a meeting between three traveling salesmen. It is far easier to schedule a meeting with two, and simplest of all to see them individually. By analogy, the cubical volume element will preferentially collapse first along one axis, undergoing a slower collapse or expansion along the other two axes. The first axis along which collapse occurs will be selected randomly if the initial deformations are random. The ensuing collapse will consequently be highly anisotropic. As both the thickness and volume of the cube decrease, the density becomes extremely high. A flat pancakelike region of high density is formed.

Evolution of voids

At first, the pancakes develop at isolated spots where the initial perturbations are largest. These flattened regions soon grow, forming thin sheets that eventually intersect. Huge cellular structures develop. The minimum dimension of a cell wall is the scale below which any preexisting small-scale structure was suppressed by radiative smoothing in the early universe. Only because of this suppression does the cellular pattern develop; without it there would be structure over all scales, and the cellular nature of the matter distribution would be much less pronounced. Numerical simulations of the collapse suggest that the universe is in this first stage today, having only recently acquired a cellular structure. In the future, as larger and larger clumps of matter form, the cellular structure may be expected to gradually disappear. It is just this intermediate stage of cosmic evolution that reflects the initial density fluctuations. The observational evidence demonstrates that the universe is neither very young nor very old from the perspective of the origin of large-scale structure.

Shocks

Hitherto, the effects of gas pressure were neglected. However, as cusplike surfaces of high density develop, the pressure builds up. The velocity of the collapsing gas soon exceeds the speed of sound. This signifies that shock fronts must develop. Behind a shock, the gas is heated to tens of millions of degrees Kelvin. Particle collisions produce prolific emission of ultraviolet and X-radiation. This energy loss by radiation has the effect of cooling the shocked gas. The cooling is greatest in the central layer of a newly formed pancake where the particle density is at its most extreme value. This cold gas layer is crucial for subsequent galaxy formation since it is unstable to fragmentation. Only cool gas is capable of forming clumps of galactic or subgalactic mass. Indeed, the typical fragment size has been estimated to be only one hundred million solar masses, comparable to the mass of a dwarf galaxy. We will see below that luminous galaxies form by the aggregation of many smaller fragments. Most of the shocked gas remains hot, however, destined to eventually become trapped in groups and clusters of galaxies. It is especially in the vast intergalactic spaces of the great galaxy clusters that large amounts of x-ray emitting gas are observed. Not all intracluster gas is primordial since the x-ray spectrum of a typical cluster shows evidence of iron and a few other heavy elements. The intergalactic gas has evidently been contaminated by ejecta from explosions of massive stars within the clusters of galaxies.

Radiation anisotropy

The pancake theory in its original form was applied to gas. It was soon realized that there was a severe problem with at least one of its predictions. The cosmic background radiation is observed to be highly uniform in intensity. Once, when all matter was ionized and the density was sufficiently high, the radiation was frequently scattered by free electrons. As we have seen, the coupling between matter and radiation only became ineffective when the electrons combined with protons into hydrogen atoms some three hundred thousand years after the "big bang." The radiation propagates freely to us after this epoch of last scattering. Any radiation temperature fluctuations must, therefore, reflect the presence of primordial inhomogeneities in the matter distribution. Astronomers have performed many searches, but failed to see any such temperature fluctuations. The best upper limit on possible fluctuations is about one-thousandth of a percent over an angular scale of six degrees of sky. This immediately tells us that the early universe was highly uniform. The fluctuations required by the pancake theory are actually incompatible with observations in a universe whose mean density is below about one-tenth of the critical value for closure. This is because there is less growth of primordial fluctuations in such models, were gravitational forces are unimportant over large scales at the present time. Hence larger initial fluctuations are required. Since such a low density represents our best estimate of the cosmological mass density, it would seem that the pancake theory is in dire straits.

Evolution

Massive neutrinos to the rescue

A neutrino rest-mass of about 0.0001 electron masses has emerged as the most likely candidate to rescue cosmic pancakes from oblivion. Neutrinos were once assumed to possess zero rest-mass as do photons, but grand unification theories have reopened the possibility that neutrinos possess mass. Indeed, the response of the particle physicist is, why not?

Experiments have tentatively measured or set limits on the rest-mass of the electron neutrino. By a curious coincidence, a neutrino mass of the magnitude that is just consistent with positive experimental evidence would have profound consequences for cosmology. The number of primordial neutrinos is predicted by the "big bang" model to be roughly comparable to the number of photons. At present, the mass-density in microwave photons is about one-hundred thousandth of the closure density. If the rest-mass energy of a neutrino were a thousand times the average energy of a photon in the microwave background, the neutrinos could dominate the mass-density at present, that is to say, they would make a greater contribution to the density of the universe than all known forms of baryonic matter. If the neutrino mass were one hundred times greater still, the mass in neutrinos would even suffice to yield a closed universe. The plausibility of such a large cosmological mass density will be discussed below.

First, we will describe the implications of a neutrino rest-mass for galaxy formation. A finite neutrino rest-mass may imply that cosmic pancakes are inevitable. Neutrinos, unlike protons or electrons, collide very rarely. Consequently, neutrinos are freely streaming, much like a hive of bees streams in all directions on suddenly being released. Any initial density fluctuation would be erased, provided it was within the distance covered by neutrinos which are moving at the speed of light. Now one consequence of neutrinos possessing mass is that below a certain energy, corresponding approximately to that of their rest-mass, the neutrino moves at a speed appropriate to its energy and slower than the speed of light, just as would any massive particle. The neutrinos slow down because of the continuing expansion of the universe. In the present epoch, free neutrinos are expected to have a speed of only six kilometers per second if their mass is 30 electron volts. This implies that the neutrinos become more and more susceptible to being trapped in density fluctuations or in galaxies when galaxies form. The maximum distance over which neutrinos were able to stream freely during the entire history of the universe amounts to about 100 million light years at present. The streaming destroys any preexisting density fluctuations, and all smaller-fluctuations were erased.

But fluctuations whose size exceeds the maximum scale over which neutrinos can stream freely not only survive, they can actually grow over a much more extended period of time than can matter fluctuations. Drag by the radiation on the ionized matter prevents growth of matter density fluctuations prior to the atomic era. Neutrinos, on the other hand, collide very rarely with photons or electrons and experience practically no drag. Consequently, the net growth of neutrino fluctuations can greatly exceed that of inhomogeneities in the old, neutrinoless theory. Thus pancakes survive and there is no violation of the near uniformity of the cosmic microwave background: this is the good news. The bad news is that the suppression of small-scale structure means that massive neutrinos, dubbed ''hot dark matter'' because they are fast-moving or ''hot'' and so suppress galaxy formation at early epochs, result in a top-down formation scenario. Galaxy clusters form first in giant pancakes that subsequently fragment into galaxies. This does not at all resemble the universe we see, where very distant and old galaxies are found, and many galaxy clusters seem dynamically young. Hence hot dark matter in the guise of neutrinos revived pancakes but failed to form galaxies in a timely manner.

Theories of supersymmetry, which pair every known particle with a hypothetical massive super-partner particle, predict that a stable, massive, weakly interacting particle should emerge from the very early universe to be the dark matter. A favorite candidate is the supersymmetric partner of the photon, the photino, a neutral particle that weighs in at 50 or more times the proton mass. Such a massive "ino" particle is slow-moving or "cold" at the onset of structure formation, hence the name "cold dark matter". A natural consequence is the bottom-up sequence of structure formation, galaxies forming first, followed later by galaxy clusters as larger and larger structures develop out of the expanding universe.

Radiation does not impede the growth of fluctuations in the cold dark matter prior to the atomic era. Consequently, one has a boost in the amplitude of dark matter fluctuations relative to the fluctuations found in a purely baryonic universe. The resulting temperature fluctuations, that correspond to the primordial fluctuations required to form galaxies by the present epoch, are correspondingly reduced in a cold-dark-matter-dominated universe by about a factor of ten.[11]

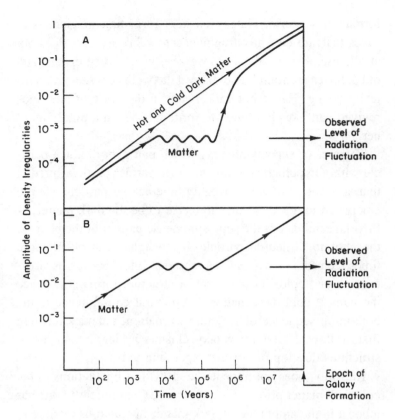

Figure 4.2 *Temperature Fluctuations*. The level of observable temperature fluctuations left in the radiation if (*top*) cold dark matter dominates the matter content of the Universe, and (*bottom*) in the absence of any non-baryonic matter.

Hidden mass

Cosmologists have a dark secret: they have not the slightest clue to the nature of the dominant mass constituent of the

Evolution

Universe. Only plausibility arguments and Occam's razor stop them from postulating the existence of vast numbers of invisible objects floating in space. For many years, the preferred candidates have alternated between black holes or degenerate black dwarf stars of the substellar variety. Remarkably, we deduce that the mass density in this invisible form of matter must exceed that in luminous material by between one and two orders of magnitude.

The units in common usage for measuring the amount of matter and luminosity in a given region are given relative to the mass and luminosity of the sun. In our solar vicinity, the amount of matter corresponds to a mass-to-light ratio of about 2. The specific type of light being measured is blue light (ideally, one would like to use bolometric or total light, but this is hard to measure). This means that the matter around us consists typically of stars of somewhat lower mass than the sun. In a great cluster of galaxies, such as Coma, the mass-to-light ratio is about 300. Part of this difference is because of the type of stars present in the cluster. Most of the cluster galaxies are ellipticals that lack young, hot, blue stars. These stars contribute little to the mass of a galaxy such as ours but dominate its light output. Thus, the mass-to-light ratio near the sun, which reflects the underlying old disk and bulge populations of stars most comparable to stars in ellipticals, is about 6. Indeed, this is confirmed by measurements of the velocities of luminous giant stars at a considerable height above the disk, which directly sample the local gravitational field. It follows that the total identifiable stellar content of the Coma cluster of galaxies amounts to only about 6/300 or 2 percent of its dynamical mass.

Furthermore, a substantial fraction, about 10 percent, of the mass of the Coma cluster has been discovered to be in the form of hot, x-ray emitting gas. Thus, the total content of the Coma cluster that is in luminous form is only 12 percent of its total mass. The remaining 88 percent of the mass of Coma is invisible, at least up to now, and its nature is a continuing source of speculation. A similar fraction of nonluminous matter appears to be present in the spiral-dominated groups of galaxies (although data on the intergalactic gas content of groups is very incomplete). These measurements both refer to scales of one or two million light years.

On a smaller scale, galaxy rotation curves sample the distribution of dark matter in the outer parts of galaxies. Studies of the rotation of our own galaxy and of other spiral galaxies have revealed the surprising result that Kepler's laws are not obeyed. The rotational velocity does not decrease with increasing distance, from the edge of the galaxy, as does, for example, the orbital speed of planets at greater and greater distances from the sun. The accepted resolution is the presence of an extended halo of dark matter.

If enough, optically invisible, mass is present, the measured rotation velocity remains constant with increasing distance from the center of the galaxy just as is observed. Dark halos appear to contain the bulk of mass of spiral galaxies. The predominance of flat rotation curves for spiral galaxies out to several times the optical radius of the galaxy indicates that the mass-to-light ratio is rising with increasing scale (see Figure 4.3). The group and cluster data suggest that the invisible mass fraction of the universe continues to rise at least until we begin surveying scales of millions of light years. Whether there is still more dark matter unclustered on larger scales is unknown. One can say that for the universe to be of critical density for it to eventually recollapse, the required amount of dark matter has a mass-to-light ratio of about 1000. Measurements of the Hubble recession velocity versus

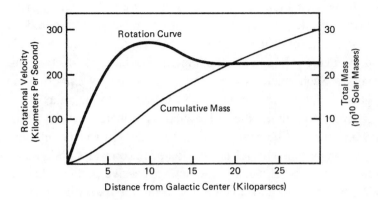

Figure 4.3 *Rotation Curve of a Galaxy.* The variation of the rotation velocity versus distance from the center of a typical spiral galaxy. At large distances the rotation velocity levels off. This implies the cumulative mass in the galaxy must continue to rise with distance from the galactic center until the boundary of the galaxy is reached.

distance relation for distant galaxies suggest that the mean density of the cosmos certainly cannot much exceed this collapse value. This critical mass-to-light ratio is only a factor of 3 larger than the Coma cluster value.

Does this signify that the universe is destined to collapse? Almost certainly not since rich clusters like the Coma cluster are rather rare objects. Far more typical is the sparse group which has a characteristic ratio of mass-to-blue-light of about 50. The large values of the mass-to-light ratio in clusters are due in part to the older and dimmer star populations, and in part to the fact that a substantial amount of matter that would perhaps have ordinarily formed spiral disks is in the intergalactic medium.

The phenomenon that is responsible for the dark matter is a universal one. Even studies of galaxy halos where little light is seen indicate that the ratio of mass-to-light must be extremely large locally, amounting to several hundred or more. The evidence seems overwhelming that exceedingly dark matter dominates the mean mass-density of the universe. Such matter exists and is dynamically important in galaxy clusters and groups, and in the halos of galaxies. It is even present in the solar vicinity, where, however, the principal contributor to the density is the old star population.

Even particle physicists have joined the quest for dark matter in the cosmos. If weakly interacting massive particles (a.k.a. WIMPs or cold dark matter) constitute the dark matter in the halo of our galaxy, the flux of these particles through the Earth is appreciable: about 10 per square meter per second. Deep underground experiments, sited to diminish possible cosmic ray contamination of the detectors, are being built to search directly for these particles by detection of the energy transferred when a WIMP collides with a target nucleus in a low-temperature crystal sensor. Such collisions are rare, but can reveal the local presence of dark matter. At the same time, if such particles as photinos exist, experiments to be performed with the next generation of particle accelerators will be capable of searching for their signature in elementary particle interactions.

Cold dark matter is also testable by simulating the cosmological process of gravitational instability and the ensuing structure formation in large computer experiments. Gravitational interactions are well understood, and the main limitation is computer power. Astrophysicists can study the evolution of some 10 million point mass particles in a cube that represents a section of the Universe of dimension 100 million light years along each side. The computer universe expands, fluctuations grow, and the output provides a three-dimensional perspective on a representative volume of the universe, containing thousands of galaxies.

Evolution

The first condensations

These speculations provide the basis for a very tentative model for the evolution of large-scale structures in the universe. The largest structures develop first, yet such agglomerations are observed to possess only a modest density enhancement over the average density. This tells us that galaxy formation must have been a relatively recent phenomenon, occurring after the universe was about one billion years old. Before then, the universe was much too dense to have developed such diffuse condensations. Astronomers have searched vigorously but with little success for positive evidence of this theory of recent galaxy formation. Should their efforts ultimately meet with failure, an alternative theory awaits their attention.

Fluctuations in the primordial *entropy,* or baryon-to-photon ratio, are not subject to any primordial smoothing in the early universe. They consist of variations in matter content within a sea of perfectly smooth primordial radiation. The radiation field is uniform and does not drag along the density inhomogeneities. The baryons are prevented from responding to the slight local excess of gravity in a density fluctuation, however, until matter becomes predominantly atomic and the radiation decouples—only then can the matter move freely. Once it does, gravitational attraction causes fluctuations to grow on scales as small as one million solar masses. These fluctuations have masses far smaller than those of luminous galaxies. The fluctuations can develop into the first condensations to collapse within a million years of the "big bang." Gravitational forces will cause the clustering of larger and larger accumulations of matter. These initial condensations are the possible building blocks from which galaxies gradually develop.

Hot particles and cold particles

Hot dark matter fails abysmally to account for galaxy formation. But massive neutrinos are not Nature's only possible contribution to the large-scale structure of the Universe. Other types of exotic particles may exist that were already cold when they formed. 'Axions' are such particles, predicted in certain theories of elementary particle structure. Small primordial black holes could also play a similar role, if produced in sufficient number. The best candidate, however, for a cold dark matter particle is the photino, or some similar particle: the stable, massive, weakly interacting relic of a bygone era of supersymmetry. Cold by virtue of its mass and slow motion the photino, like other cold particles, clusters on all dimensions from stellar to galactic, and these fluctuations in density will survive. For much of cosmic history, the gravitational influence of these particles is negligible compared to that of the radiation. Fluctuation growth only begins in earnest once the universe is matter-dominated, some 10^4 years after the initial singularity.

A unique feature of the subsequent clustering is that it is likely to occur with comparable strength simultaneously as galactic masses of all scales collapse. Larger scales undergo less growth within the horizon and their collapse is consequently delayed. Hence galaxy clusters form later: there is a natural hierarchy from bottom to top as large-scale structure develops. Perhaps there is some hot dark matter, too: a mixture of two-thirds cold and one-third hot (corresponding to choosing a mass of 8 electron volts rather than 30 electron volts for the neutrino mass) produces a more attractive mixture of large-scale pancakes and small-scale galaxy formation than cold dark matter alone.

Galactic building-blocks

Whether one believes in the existence of primordial curvature wrinkles or in entropy fluctuations, and whether large-scale structure forms at the same time as small-scale structure, or long afterwards, the implications for galaxy formation arising from these two pictures are rather similar. Pancake fragments or individual clustering clouds contain some millions, or even hundreds of millions, of solar masses. The characteristic masses are far below those of the average galaxy. The aggregation process by which these building blocks assemble into galaxies must be responsible for many of the characteristic properties of galaxies. Just as adult behavior patterns and personalities reflect environmental influences during infancy, so can we also hope to learn about the process of galaxy formation by studying the morphology of the galaxies. Galaxies are gigantic fossils.

Because of the vast distances between stars, dynamic interactions between them now proceed at a negligible rate. But during the epoch of galaxy formation, things were very different. Many of the stars we see now were unborn, just components of gas clouds vigorously interacting. Stars are born, evolve, and shed enriched debris that is in turn used to manufacture the next generation of stars. Once the bulk of star formation ceases, the distribution of heavy elements that we see in old stars provides evidence of enrichment that occurred long ago. Even after stars form, there can be a great deal of mixing of stars and orbits in a system that is undergoing overall collapse. It is these types of processes that we must consider if we are to understand how galaxies acquire their characteristic features. First, however, it is helpful to review the important properties of the objects we are seeking to explain.

Spirals

A galaxy such as our own Milky Way system consists of a hundred billion or more stars moving in a perpetual dance orchestrated by gravity. The time for the sun to complete an orbit of the Milky Way is about two hundred million terrestrial years—a period that may be described as one galactic year. Our galaxy is about fifty galactic years old and now in the prime of its life. The stars in the thin disk of our galaxy move in nearly circular orbits around the galactic center. In addition, a second population of stars moves on eccentric orbits, often highly inclined to the disk. These constitute the galactic "bulge." When spiral galaxies are viewed edge on, the bulge usually appears as a luminescent ball of light on which a thin dark band has been superimposed. This band is due to interstellar gas and absorbing dust which is concentrated into the disk and silhouetted against the bright central bulge.

Evolution

A small percentage of the mass of a typical spiral galaxy like our own is usually in the form of interstellar gas and dust particles. It is this diffuse matter that provides the birthplace for stars and that reiterates how immensely old the bulk of the stars must be. For only in some remote past epoch could there have been enough diffuse nonstellar material from which to create all the stars (see Figure 4.4). The interstellar matter tends to be compressed into a spiral pattern in the disk of the galaxy. The spiral is created by a cosmic traffic jam. As clouds orbit the galaxy, they accumulate in certain regions, and the congestion pattern slowly spirals outward as the galaxy rotates. In these spiral arms, the new stars form continually. We recognize the spiral pattern on photographs primarily because of the presence of newly formed and luminous blue stars.

Ellipticals

Some galaxies do not possess a stellar disk at all. These elliptical galaxies are all bulge, consisting of old red stars. Little interstellar matter is seen in ellipticals, and the star formation process is now extinct there. Elliptical galaxies are gigantic fossils revealing traces of an active youth tens of billions of years ago. What cosmic process could have intervened to halt star formation and remove the interstellar matter in ellipticals? A crucial clue comes from the environment. Ellipticals predominate in the richest and largest galaxy clusters. In regions where the population of galaxies is densest, one finds that ellipticals greatly outnumber spirals. In regions only sparsely populated by galaxies, the spiral galaxies

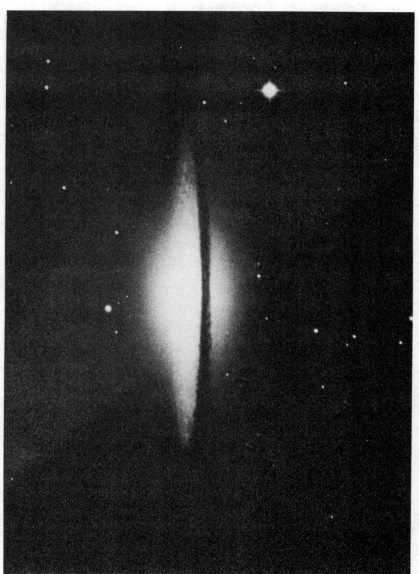

Figure 4.4 *Galaxy Types.* A selection of galaxy types illustrating the spiral galaxies Messier 104 in Virgo and Messier 81 in Ursa Major, and the elliptical galaxy Messier 87 in the Virgo cluster.

Figure 4.4 (continued)

are the dominant variety. It seems, then, that environmental effects must play a vital role in determining the morphology of galaxies.

Galactic spheroids

The ellipticals are spheroids. The spirals are a hybrid mixture of spheroid and disk, and other, irregular, galaxies appear to lack any spheroidal component at all. Spheroids exclusively contain old stars, whereas the disks are often regions of active star formation. It is to the spheroids we must look for fossilized traces of galactic birth. Stars in galactic spheroids are practically as old as the universe and formed during, or very shortly after, the epoch of galaxy formation.

One universal property of spheroids is the smoothness of the light distribution. Whether one looks at a giant elliptical with a hundred billion stars or a dwarf spheroidal galaxy with only a million stars, the light intensity decreases smoothly with radial distance from the center. The light intensity profile of spheroids obeys a universal form, named originally, after Edwin Hubble, and later modified by Gerard de Vaucouleurs. Double the radius, and the light intensity decreases approximately by a factor of four. This systematic decay of brightness continues without limit until the spheroid is lost in the faint night sky glow. Somehow, the galactic birth process gave rise to the Hubble–de Vaucouleurs profile.

The central regions of an elliptical galaxy are exceedingly bright. Stars are packed together by the million in each cubic light year. Of course, we cannot identify individual stars, but their collective presence can be inferred from the spectrum of the emitted light. Individual types of stars can be recognized because of peculiar spectral features or lines due to the presence of various heavy elements. These spectral lines are blurred because of Doppler shifts caused by the to-and-fro motions of the stars, but the lines are still recognizable. Indeed, from measuring the line widths and average line wavelengths, some vital stellar statistics can be inferred. We learn the description of stellar velocities and the heavy element content of the stars. We can tell whether or not the galaxy is in a state of rotation. We can also measure the random component of stellar motion, which is conceptually similar to the temperature of a gas (only here the stars play the role of molecules).

From spectroscopic measurements of elliptical galaxies, we have learned some surprising facts. Luminous ellipticals do not seem to be rotating at a significant rate. That is to say, centrifugal forces are not responsible for supporting stars against gravity. On the other hand, stellar disks are clearly supported by rotation. What, then, is enabling the stars in a spheroid to avoid collapse into the galactic center? Simply, energy. The stars that occupy the spheroid are moving in random directions at sufficient speed to maintain a stable configuration. Random stellar motions are also responsible for the roundness of the sphcroids. Often, stars in the inner regions appear to have a higher content of heavy elements than the more remote stars. Metallic ions in stellar atmospheres make them more opaque and tend to depress the effective temperature of the star. This effect manifests itself as a reddening of the central light from galactic bulges.

Evolution

Environmental impact

Galaxies undoubtedly have some characteristics that are influenced by their local environment. In the regions most tightly crowded with galaxies, ellipticals predominate, whereas in regions where the galaxy density is low, we find spirals almost exclusively. Galaxies also span a wide range in luminosity. The giants usually contain more rapidly moving stars and a higher metal content than do the dwarfs. Such tendencies echo a capitalistic society where the rich grow ever richer at the expense of the poor. Galaxies as we will see, also grow by consuming lesser systems.

Colliding clouds

Model-building can serve a useful function. Whether in teaching a child or in designing an airplane, the experience gained is an invaluable guide to reality. The astrophysicist has little choice. Since the stars are beyond his grasp, building models is his only recourse. His tools are simple, pen, paper, and computer (perhaps also a large waste-paper basket!), but his horizons are unbounded—the universe awaits his pleasure. Let us, then, stir the cauldron and concoct a galaxy.

The basic ingredients are clouds of gas, with masses in the range 10^6 to 10^8 suns. These clouds lead an isolated and uneventful existence until a close encounter occurs between two of them. Such encounters occur rapidly once cloud accumulation begins in earnest. A typical collision occurs at a velocity that is large compared to the characteristic notion of gas atoms in an individual cloud. Just as in a high speed collision between automobiles, each cloud is strongly compressed, as a shock wave travels through them. The compression raises the gas density, but only heats the gas in a thin layer. The average temperature of the cloud does not increase because the gas is able to radiate away most of the excess heat imparted by the compression. The sudden rise in pressure causes the density to rise drastically as the gas cools down after the passage of a strong shock. A cold, thin, dense layer develops in which the role of gravity is enhanced. Studies of compressed clouds suggest they will become unstable, collapse, and eventually break apart. Finally, stars form out of the self-gravitating gas fragments. A similar sequence of events is observed in galactic spiral arms. The collision rate between interstellar gas clouds is enhanced as cloud orbit after orbit becomes ensnarled in the deepening gravitational potential well that demarcates the spiral pattern. The result is a virtual epidemic in formation of short-lived massive stars, apparently triggered by interstellar gas cloud collisions.

Separation of the stars

Once stars are formed, a parting occurs with the surrounding gas. Rather as a speeding bullet conserves its momentum and ploughs on through the atmosphere, the young stars tend to maintain whatever kinetic energy and momentum they acquired at birth. Meanwhile, the parent gas clouds recklessly continue their careers of dissipation, colliding with other clouds and losing orbital energy. More stars may form in ensuing cloud collisions, but the fate of the gas is sealed. Having lost much of its initial energy and angular momentum, it falls towards the galactic center, the eventual repository for all remaining gas. A strong concentration of stars grows towards the galactic center, as more and more stars are formed by the rapidly eroding clouds. The central bulge builds up, and the process stops once the reservoir of clouds is exhausted. Residual debris collects in the nucleus of the galaxy, possibly collapsing to form a massive central black hole.

Disk formation

Meanwhile, clouds with sufficient angular momentum in their orbits will remain in the outer regions of a protogalaxy where they rarely encounter other clouds. Such clouds are likely to survive. Since the gas retains angular momentum, even as the clouds are disrupted by collisions or by star formation, the gradual erosion of surviving clouds must result in the eventual formation of a disk of stars around, and even embedded within, the spheroidal component. The overall collapse and loss of energy in random cloud motions conspire to give the stars nearly circular orbits and create a thin disk. Disks consist mostly of material that has undergone ongoing strong dissipation. The stars have lost practically all of their component of motion out of the plane of the disk. In the spheroidal system of stars, dissipation appears to have terminated much earlier, leaving a strong residue of stellar motions in random directions. Close inspection of spiral galaxies often reveals an old, thick disk of stars, intermediate in its properties between the spheroid and the thin disk of more massive, younger stars. These young stars are found in a layer no thicker than 300 light years in which all of the interstellar matter and present-day star formation activity occurs.

Mergers

Simple laws of physics governed the origin of galaxies. We can understand many of the gross characteristics of galaxies by means of our simple building-block model. Some galaxies are all spheroid, some mostly disk. We interpret this in terms of environment. An isolated cloud aggregate will slowly develop a more and more pronounced disk. Contrast this with a cloud aggregate that merges with other nearby systems of clouds. Any memory of a preexisting disk will tend to be lost during the randomization of orbits that occurs during the merger. One can then readily speculate why ellipticals, pure spheroids of stars, predominate in dense regions, such as rich clusters of galaxies. Collisions are a relatively frequent occurrence there. Nowadays, a collision between two galaxies in a rich cluster has no more impact than the passage of two ships in the night—individual stars do not collide. Only during early galactic evolution, when the galaxies consisted predominantly of gas clouds, would the interaction have been much more dramatic (see Figure 4.5).

If the collision was a gentle one, the systems would merge, but if it occurred at high velocity, as large as that found in a great cluster, the gas clouds would undergo an orgy of destruction, heating to thousands of millions of degrees and

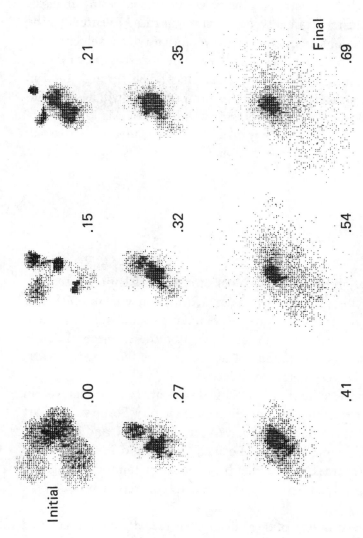

.00   Initial

.27

.41

.15

.32

.54

.21

.35

.69   Final

Figure 4.5 *An Encounter Between Galaxies.* Computer simulation of the collapse of 5,000 mass points. Initial irregularities in this time sequence evolve into a smooth, centrally condensed distribution resembling an elliptical galaxy.

dispersing throughout intergalactic space. When a great cluster first develops out of the expanding universe, the relative velocities between galaxies are small. Gathering up mass, the cluster gradually aggregates galaxies. Eventually, the gravitational acceleration grows so that typical velocities between protogalaxies are high enough to destroy any remaining clouds. Only after a galaxy has fallen through the full gravitational potential of the forming cluster will it attain full speed. This marks the end of the protogalactic phase of evolution. The building blocks are consumed or destroyed; only mature galaxies remain.

The seven dwarfs

But are the building blocks for galaxies entirely used up? It seems unlikely that galaxy formation was so efficient a process, and some left-over remnants should still be around today. Dwarf galaxies could well be these objects. Loosely bound agglomerations of as few as a million stars, dwarfs may be very common indeed.

The least luminous of these dwarfs are so faint that we can only detect them in our immediate vicinity. Seven such dwarf satellites of our galaxy have been discovered, and bear such exotic names as Draco, Fornax, Carina, and Sculptor. These dwarfs are so diffuse that they appear in imminent danger of being ripped apart by the tidal force exerted by our own galaxy.

One property of these dwarfs has recently been discovered that promises to be important for discerning the nature of the dark matter in the universe. It appears that several of the dwarfs, and in particular the Draco System, may themselves contain a considerable amount of dark matter. Massive neutrinos cannot account for the dwarf masses, since the neutrinos cannot be packed sufficiently closely together. The exclusion principle of quantum mechanics limits the number of quantum states accessible to the neutrinos: when these are filled, the neutrino gas is said to be degenerate and resists any further compression. However, photinos are one exotic type of elementary particle that we have previously encountered, which, unlike neutrinos, are not subject to this limitation. Photinos may be packed together as closely as gravity requires. Should the high ratio of about thirty solar masses per unit of solar luminosity be confirmed for Draco and other dwarfs, it could favor some cold particle such as the photino as the best candidate for dark matter.

Are there any antigalaxies?

We believe the universe began in a symmetric state, there being precisely equal amounts of matter and antimatter initially. Yet, in our immediate locale, antimatter is absent. This preponderance of matter over antimatter is a consequence of the symmetry breakdown at the epoch of grand unification, 10^{-35} second after the "big bang." The orthodox view is that the peculiar imbalance between matter and antimatter we see was frozen into the universe everywhere at that early moment. Conceivably, however, the transition could have occurred in a totally chaotic fashion, leaving vast domains of matter surrounded by regions of antimatter, the relative sizes varying randomly from region to region of the universe. The choice between ending up as matter or as antimatter depends on the orientation of certain quantum fields in a particular region. If the orientations are truly random, like flipping an unbiased coin, then averaged over the whole universe any bias for either matter or antimatter will be ironed out. How large could these domains of matter be? Certainly, no smaller than our galaxy, for interstellar matter pervades the Milky Way. Even if a tenth of 1 percent of material were antimatter in our galaxy the ensuing annihilation from the mixture of antimatter and matter would lead to our being bathed in gamma rays to an extent far in excess of that observed.

It is possible for galaxies and even clusters of antimatter to exist. Optically, the antigalaxies and anticlusters would be completely indistinguishable from ordinary galaxies and clusters. However, intergalactic gas pervades much of space, and it would be extremely difficult to avoid mixing matter with antimatter. Only by arranging for our local domain of matter to be a considerable fraction of the entire observable universe could one suppress the radiation caused by annihilation to a level below the observational limits. This seems so artificial and improbable an arrangement that, most likely, there are no antigalaxies, and certainly, there are none close by.

The biggest structures in the universe

It has been known for many years that there is a significant excess of nearby galaxies above the average background distribution in a region that extends ninety degrees or more across the sky near the north galactic pole. This excess has been dubbed the Local supercluster, and it contains the Virgo cluster along with numerous lesser groups of galaxies. The Local Group, of which our own Milky Way and its neighbor the Andromeda galaxy are the most prominent members, lies on the outskirts of the Local supercluster, which extends for some 100 million light years. Detailed surveys of the Local supercluster find that it consists of two distinct components: a flattened layer containing 60 percent of the luminous galaxies, and a surrounding heterogeneous halo containing the remainder. Almost all of the halo volume is empty, with most galaxies in a small number of diffuse clouds.

While the Local supercluster is one that has received the most attention because of its proximity, many other superclusters have been identified. Superclusters often consist of two or three rich galaxy clusters, each of which may contain a thousand or more bright galaxies within a volume of a hundred cubic megaparsecs. The clusters are linked by filaments and chains of luminous galaxies, so that superclusters often have an elongated appearance. Deep red shift surveys reveal that much of space is devoid of bright galaxies. The galaxies are mostly concentrated in sheet-like and filamentary ridges, the most prominent of which are superclusters. Figure 4.6 illustrates a wedge of the sky containing the supercluster around the Coma cluster of galaxies.

Nuclear reactions in stars

In our quest for evolution, from the largest self-gravitating structures we now turn to the smallest. The universe ac-

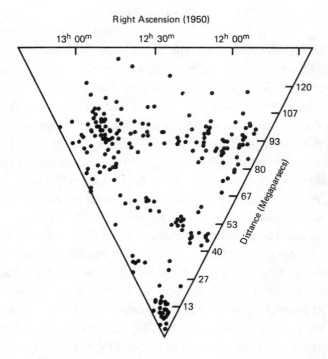

Figure 4.6 *The Coma Supercluster.* The distribution of galaxies around the Coma cluster. The locations of all galaxies are plotted within a cone subtending an angle of about 20 degrees.

quired more and more disorder on large scales while complexity and order developed on small scales. The net amount of disorder or entropy in the universe is set by the photon—to baryon ratio: as we have seen, this is dominated by the cosmic background radiation and does not increase significantly as stars form and die. But on small scales, entropy undoubtedly decreases as matter rearranges itself to form highly ordered living systems. One of the keys to this interplay between large and small scales, disorder and complexity, lies in chemical evolution. Were it not for the alchemy in stellar interiors, the universe would contain only hydrogen and helium. Solid planets could not condense; life would not exist. The first stars fused protons into helium nuclei. The slight reduction in mass when a helium nucleus forms from four protons provides the source of energy. As the stellar core exhausts its hydrogen fuel supply by burning it into helium, it shrinks under the relentless tug of gravity and heats up. Because a helium nucleus has the electrical charge of two protons, it takes more energy to overcome the electrical repulsion between helium nuclei and allow their fusion. Once the core has heated sufficiently, three helium nuclei are fused into a carbon nucleus, which weighs fractionally less than the helium nuclei. This slight mass deficit provides a renewed source of energy.

Supernova

A luminous star is a great ball of dismembered atoms, or what physicists call an ionized gas or plasma. Such stars are extremely hot and blue in appearance. Cooler stars are reddish, and while still predominantly gaseous, contain atoms and molecules rather than freely charged particles. The enormous force of gravity compresses the core of a star to such an extent that were it not for the pressure generated by the nuclear reactions ignited in its interior, the center would solidify. Indeed, old stars that have exhausted all the supplies of nuclear fuel within their interior collapse to become small, solid "dwarf" stars—just huge chunks of cooling carbon or iron.

The nuclear reactions continue, with heavier and heavier nuclei being synthesized until iron is produced by stars of about ten solar masses or more. The less massive stars end up with cores of carbon or oxygen. We call these white dwarfs. [12]

Iron is the ultimate slag heap of the universe, for no more energy can be extracted from fusion to heavier nuclei. Radioactive decays or fusion of very heavy nuclei are possible, but enormous amounts of energy would first be needed to make these nuclei. Once an iron core forms, the star has attained its final destiny. The exhaustion of the internal energy supply means that gravitational collapse is now unimpeded, and the collapse of the core releases enough heat to eject the outer layers of the star. This is a supernova explosion. The core shrinks into an incredibly compact object as dense as an atomic nucleus. Indeed, the entire core is practically a giant nucleus. We call this a *neutron star*. Meanwhile, the outer layers of the star, rich in the heavy elements produced by thermonuclear fusion, are dramatically ejected into the surrounding interstellar medium. The final testament of a massive star contains an impulsive bequest: its ashes are the future seeds of life.

Origin of heavy elements

The first stars spew out carbon and oxygen, silicon and iron. The stellar ejecta stir up the surrounding gas and become well-mixed. A new generation of stars eventually is born. These stars already bear the imprint of the past by containing heavy elements in modest amounts. The pristine interstellar matter, initially devoid of any heavy elements, becomes contaminated by the debris from evolved stars. And so it continues. Star birth followed by star death, and gradual enrichment of the matter out of which new stars form by heavier elements. But some of the details of this birth and death process are obscure. We see very few stars that are unenriched, survivors from the earliest generations of star formation. Perhaps, in order to have seeded the galaxy, primordial gas clouds must preferentially have formed massive short-lived stars rather than long-lived stars of solar mass that would still be visible. We can speculate, but we do not really understand how this happened. But then we do not really understand the details of how stars form now in nearby dense interstellar clouds either.

The oldest stars

Just as a random selection of people would be found to possess a wide spectrum of ages, so the stars around us vary greatly in maturity. Stellar ages are inferred from the rate at which a star consumes nuclear fuel in order to shine steadily in its most long-lived phase, that of hydrogen burning. The more massive stars are profligate users of nuclear fuel. Consequently, they are relatively short-lived. Their huge mass and gravity means the central pressures and maximum temperatures are much higher than within their lighter counterparts, and so nuclear reactions proceed much more rapidly at their cores. They live faster and die younger.

Look in the belt of Orion, and there are stars born only a million years ago, mere infants by astronomical standards.

By contrast, the sun is about five billion years old, and at least another five billion years will elapse before the infirmities of old age, instigated by the exhaustion of its hydrogen fuel reserves, begin to have catastrophic consequences for life on earth. The Orion stars are grouped into associations of young stars that were born more or less simultaneously. Other, older groupings of stars are also known. The most beautiful and the most ancient are the globular star clusters. Indeed, these are the oldest aggregates of stars in our galaxy having formed nearly fifteen billion years ago.

The spheroid or halo of our galaxy contains these clusters of stars. Most of these stars are very deficient in carbon and oxygen compared with the sun. Observations indicate that the halo stars formed first about fifteen billion years ago; their ejecta enriched the remaining gas clouds, which lost energy as collisions occurred and gradually settled into the disk. The oldest disk stars are about ten billion years old, and our sun formed a little later, some five billion years ago.

Are stars inevitable?

A collapsing gas cloud remains cool as its density rises because of the extreme efficiency with which atomic collisions can radiate energy away. Thermal energy and gas pressure provide little protection against the destabilizing effect of gravity. One cannot avoid fragmentation of the entire cloud. The first fragments themselves are also unstable and break up into more generations of subfragments. There is much uncertainty in estimates of the minimum fragment mass that finally results, but it seems likely to be of stellar size. The increasing density eventually results in the fragments becoming sufficiently opaque that they heat up and become supported against gravity by pressure forces. One now has a collection of protostellar objects that are still slowly contracting. They eventually form stars when their cores reach a high enough temperature to trigger thermonuclear reactions. Similar physical effects operate both in the final stages of the evolution of a collapsing cloud and during the more quiescent stages of stellar evolution. Indeed, this is why there is only a relatively small range of mass spanned by nuclear-fueled stars. The least massive stars are one-tenth the mass of the sun, while the most massive stars are about one hundred solar masses.

Interstellar grains

The interstellar medium appears very prominently on photographs of the Milky Way either through regions of extreme obscuration or those of incandescent nebulosity. The dark obscuring clouds contain tiny particles of dust that absorb and scatter the light shining on them from background stars. Distant stars are reddened because of the scattering of the shorter wavelength light by interstellar dust particles (a similar effect in the terrestrial atmosphere is responsible for the reddening of the sky at sunset). The gas will be ionized if a hot star is present and will emit light that is detectable on a deep photograph. Hot stars illuminate their surroundings and provide us with a glimpse into the otherwise dark and quiescent clouds where stars are born.

Most of the elements heavier than helium condense into

solid particles at the freezing temperature of interstellar space, some ten degrees Celsius above absolute zero. Some of these particles, smaller than grains of sand, drift together and coagulate into larger particles. The particles lie in great clouds of gas that orbit around the plane of the galaxy. The clouds occasionally collide and coalesce. Eventually, the effects of the self-gravity of the cloud will be so enhanced that collapse commences. The densest regions continue to collapse. Once collapse is initiated, it is very hard to resist the pull of gravity.

Protosolar nebula

One part of one cloud collapsed 4.6 billion years ago. The dense center grew denser and hotter while gas swirled around in a fat disk. Most of the angular momentum remained in the disk as the central protosun developed. The grains fell like hailstones through the gas into the midplane where they swept up more particles. The volatile snow-like matter readily coalesced, and the largest grains grew larger. The solar system was under way. Within tens of millions of years, most members of asteroid-like planetesimals had formed. Gravity aided the capture of more and more rocks until a planetary system developed. Each planet devoured practically everything within its orbital path. The volatiles condensed farthest from the sun, and the giant outer planets captured vast amounts of hydrogen and helium, while the inner planets retained most of the heavier refractory grains. One of the results was the planet Earth, and this is where our story of cosmic evolution ceases (see Figure 4.7).

Evolution

Our journey has been lengthy. It has led us from an observable universe containing only one-millionth of a gram to one containing great superclusters of galaxies. We have viewed cosmic evolution over time-scales as short as 10^{-43} second up to tens of billions of years. The beginnings are buried and unobservable. But it is to these beginnings we must return if we are to ascertain whether our elaborately detailed theory is valid or merely a modern mythology, one version of one tale among many others.

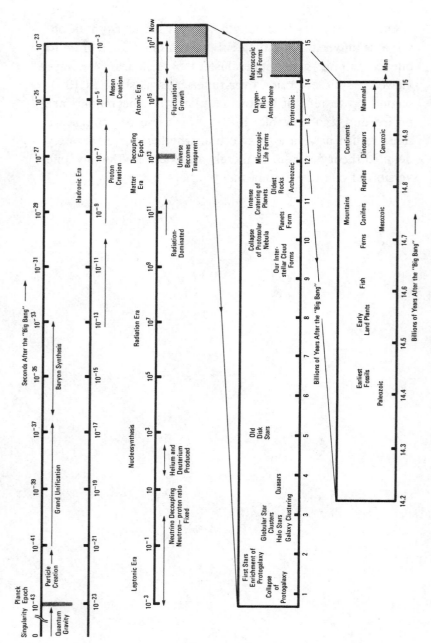

Figure 4.7 *Panorama of the Universe's History.*

5

Chaos to Cosmos

Our imaginary journey from the singular origin of space and time to the intricate luminous structures that now extend beyond the range of our largest telescopes has taught us an important lesson. The present make-up and appearance of the universe is inextricably linked to conditions that existed in the remote past before the appearance of stars, galaxies, and astronomers. Things are now as they are because they were as they were. This direct causal link which exists between the past and the present provokes us to pause and inquire just how different the present-day universe could have been. An adequate answer to such a question requires us to decide which aspects of the universe's structure are inevitable, which are perhaps bound up with inflexible conservation laws of nature, and which arise as just one of many equally likely possibilities. Which properties of the cosmos could now be different if the initial conditions of the universal expansion had somehow been radically altered? With this question in mind, we now probe in a little more detail several special properties of the observed universe that hint at unusual events in the very earliest moments of cosmic time. Such events motivate much of the contemporary research into the history of the universe because they provide unanswered, and perhaps even unanswerable, questions.

Symmetry

Pick up a tennis ball, turn it around and examine it. It is extremely smooth and spherical. Any deviations from perfect spherical symmetry are slight and barely perceptible to the eye. Yet our observations of the microwave background radiation, that is incident on the earth from all directions in the sky, show that the entire observable universe appears more symmetrical even than this. The relative difference in the expansion rate of the universe measured between any two directions around the sky must be less than a tenth of one percent. Intuitively, this implies that the universe expands like a perfectly spherical ball to extremely high precision when viewed from any cosmic point. We say that the cosmological expansion is *isotropic;* that is, it displays no directional bias.

Imagine a cosmic observer who moves around the universe surveying vast aggregates of galaxies. He would detect a striking systematic trend in their spatial distribution. As larger and larger regions of the universe, containing more and more galaxies, are sampled, a comparison of the total amount of matter measured in different regions of identical volume leads to an increasingly similar answer. Over a cubic region, three billion light years in dimension, the samples differ in mass by less than a percent. These observations are telling us that the universe becomes increasingly uniform as we examine larger and larger pieces of it. The lumps and bumps that appear so noticeable on small-scales—whether planets, stars, or galaxies—gradually average out when viewed overall. We say that the universe appears to be spatially homogeneous when surveyed over regions of sufficiently large extent.

The high degree of isotropy and the homogeneity in the universe are remarkable properties reminiscent of the perfection sought by the ancient Greek cosmologists. Indeed, until the mid-1960s, cosmologists viewed the universe as a smooth, expanding substratum of material on which there appeared small condensations in the form of galaxies and stars. They took the existence of the symmetrical, expanding substratum for granted and attempted to explain how the small condensations could have arisen by the process of gravitational instability. While Einstein had originally postulated these idealizations out of a blind faith in the simplicity of nature, the astronomical evidence for them has continued to mount ever since Hubble first counted galaxies in different directions in the sky. It has since withstood scrutiny by instruments of ever-increasing sensitivity.

A dramatic discovery led to a radical overhaul in cosmological thinking in 1965 when the first measurements of the cosmic microwave radiation's temperature isotropy were made at Princeton University. The results were so striking that doubts began to surface as to whether it was really the small nonuniformities in the universe that were most in need of an explanation. Perhaps the extraordinary regularity of the overall expansion was the major mystery. After all, irregularity can occur in many different ways, but regularity in only one. Disorder is much more probable than a state of regularity that requires a very special configuration. Why should the universe be so symmetrical today?

Primeval chaos

This question led to the formulation of a modern version of a hypothesis that has its roots in ancient times. The precise opposite of an orderly, smooth beginning to the universe was an expansion out of complete chaos. The goal of the school of, so called, "chaotic cosmologists" was to provide an explanation for the observed regularity of the universe in a manner that made no appeal to particular initial conditions fixed at the unknown beginning of the expansion history. The originator of this approach, the American physicist, Charles Misner, wanted to show that no matter how chaotically the universe began its expansion, given enough time, it would inevitably expand and cool towards a state of quiescence and isotropy. The chaos would have been eradicated by frictional processes during the early history of the universe. Instead of merely assuming the regularity was part of the initial conditions, Misner wanted to show that it could be explained without recourse to any special assumptions about how the universe began.

A little illustration is useful. Imagine you station one of your friends on a cliff top, armed with a collection of stones, while you sit blindfolded on the ground beneath. You ask your friend to begin tossing the stones to the ground far below in any manner he chooses—some slowly, some rapidly. Now, still blindfolded, could you tell the speed at which the stones hit the ground (hopefully) beside you? At first, you might be tempted to say, "Of course not. If they are thrown rapidly, they will hit the beach hard, but as I can't see *how* they are being thrown I can't possibly predict the speed of impact." But this is not correct.

If the cliff is high enough, you can indeed predict the impact speed without seeing how hard your friend decides to throw the stones. The reason is that bodies falling in air are very quickly accelerated under gravity to a particular, or "terminal," velocity that is attained when the accelerating downward force of gravity is exactly counterbalanced by the decelerating resistance of air. When these opposing forces become equal, no net force acts on the body and, like all bodies acted upon by no force, a stone would continue to move at a constant speed. This speed is called the 'terminal velocity' and is determined by the ratio of the acceleration due to gravity and the air resistance. We expect the stones to hit the ground with this terminal velocity if they fall for long enough, no matter how they were first thrown from the cliff top. If they were flung extremely hard, the air resistance would at first dominate to decelerate them down to the terminal velocity. If they were simply dropped from rest, gravity would dominate and accelerate them up to the terminal velocity. This also means, of course, that, if we wanted to know how hard a stone was thrown, we would be stymied because it loses memory of its initial speed and hits the ground at the terminal speed regardless of how it started off.

The philosophy of the chaotic cosmologists is completely analogous. They imagined that no matter how dissimilar the "big bang" was from place to place, and how chaotic and anisotropic its initial expansion, given enough time, the more rapidly expanding regions would be slowed down by friction until different parts of the universe were expanding at the same terminal rate. Given this hypothesis, present-day observers should always see a universe that is smooth and isotropic because they live a long time after the inferno of the "big bang." Clearly, this is a very appealing approach to the cosmological problem. It renders the unanswerable question, what was the beginning of the universe like?, quite irrelevant for an understanding of its present structure. The present observation of regularity could equally well be the consequence of primordial chaos as of primordial order so long as frictional processes work efficiently in the distant past.

There are indeed physical processes that can create significant frictional drag in the early universe. In particular, when the universe is just a second old, prior to the nucleosynthesis of the lightest elements that was described at the end of chapter 3, the weakly interacting neutrinos are just ceasing to interact with other particles. The average distance that a neutrino has to travel before colliding with another particle becomes greater than the distance a neutrino is able to travel in the whole age of the universe up until that moment. That is to say, practically all collisions involving neutrinos simply cease. But, just as the neutrinos are on the verge of slipping into isolation from the rest of matter, they enjoy a brief period when they just manage to interact occasionally but need to traverse almost the entire observable universe before doing so. At this moment, they provide a significant frictional influence. If they originate in one place where the density of material or the rate of cosmic expansion far exceeds the average, they will be able to traverse the whole universe before parting with their energy excess to a more slowly moving particle elsewhere where conditions are more rarefied. In this way, large disparities in the distribution of matter can be erased, it can be ensured that the universe expands at the same speed in different directions, and there can arise a rapid equalization of the density and temperature from place to place; this transport of energy by the essentially massless neutrinos is dubbed neutrino viscosity.

It is amusing to notice the analogy here between explanations for the large-scale regularity of the universe and a more familiar argument about the nature of intelligence known as "the IQ debate." On the one hand, it is claimed that the principal contribution to intelligence is via genetic factors inherited at birth with subsequent environmental influences playing a negligible role. On the other hand, it is argued that the initial genetic make-up is largely irrelevant, and it is the environment that plays the major role in developing a person's intellectual capability. The chaotic cosmology theory is analogous to the second of these options. It asserts that the observed large-scale structure of the universe is created principally by its behavior during the early expansion history, in other words, its character is due to all the interactions that occurred and the physical processes that inevitably arose there. The alternative and more traditional view assumes that the universe began in a state of high symmetry with some small fluctuations that gradually developed with the passage of time. The present regularity in this quiescent cosmological scenario is directly attributable to the starting conditions "genetic" make-up. The evolutionary history plays a negligible role.

There is a lot to be said for both the quiescent and the chaotic cosmological hypotheses. Certainly, chaotic cosmology appears to have the greater explanatory power and promises to explain many observations with a smaller investment of special assumptions than the quiescent model which does not seem to be really explaining anything. It is merely saying the universe is regular today because it was regular yesterday. However, the price the chaotic cosmology hypothesis has to pay for this advantage is computational complexity. In order to prove it viable, one would have to study all the conceivable starting conditions for the universe and check that each and every class of them ended up becoming regular as the universe expanded. This is difficult to do. It is much easier to falsify it of course; a class of initial conditions that evolves towards irregularity will suffice. By contrast the quiescent cosmological scenario makes a very specific assumption of simplicity, the consequences of which can be readily developed.

Demise of chaos

Let us suppose, for the sake of argument, that the chaotic cosmology hypothesis is accepted, and let us see what cosmologists discover about its implications when it is examined more carefully. Using the example of the falling stones, we recall that the stones reached their terminal velocity only if they fell for long enough. We also learned that the *harder* our friend threw the stone initially, the *longer* it would take to lose memory of its starting speed and settle down to its terminal velocity. If the cliff were not very high, it would, in principle, be possible to project a stone from the top with such enormous force that it would not slow to anywhere near its terminal velocity by the time it hit the ground below. Clearly, we would not be able to predict the impact speed of this stone because it would not have washed away all memory of its initial speed by the time it reached us.

Likewise, it is always possible to conceive of starting conditions for the cosmological expansion that are *so* chaotic and so different from the regular terminal state of isotropy that physical processes could not work quickly enough to make the expansion as regular as it is observed today. At first, it was hoped that these counterexamples would not signal the death knell of the chaotic cosmology theory because these ultra-chaotic models might be very special, as special in their own way as those with precisely regular initial conditions. Suppose the surface of this page charted all the possible starting conditions of the universe. We could color red all those points that lead to an extremely isotropic universe like our own by the present-day and then color blue all those that did not. Every point on the page would be colored either red or blue. If we found the page to be colored entirely red, the chaotic cosmology hypothesis would appear very sound, but we would really be equally happy if the only blue colorations were infinitesimal isolated points. No matter how many of them there were, so long as they did not combine to form *finite* blue areas, we would be content that almost every set of starting conditions led to regularity. If a single blue point is isolated, any change of position away from it, no matter how small, lands you on a red point. Thus the initial expansion of the universe would have to have been very specially arranged in order to be represented by one of the blue points. The points lying in finite regions of the same color do not need to be hit so precisely as a near miss yields another model with very similar properties.

Now it turns out that there are many blue points on the page representing universes that do *not* become regular by the present day. Even worse, some do not approach a regular terminal state no matter how long they expand for; they just keep getting more and more irregular. Furthermore, not only are the blue regions finite but it is the red points that are isolated and improbable.

Quite the opposite of the chaotic cosmology theory appears to be the case according to our existing mathematical theory of an expanding universe; almost every set of physically realistic starting conditions leads ultimately to a universe that is extremely anisotropic and nonuniform, quite unlike our own. This suggests that for some reason, as yet unrevealed, the initial conditions of our universe belong to a unique set of very special red points—those which begin expanding in a highly symmetrical fashion. Despite this blow to its aims, the chaotic cosmology theory serves an important function. It highlights a number of otherwise neglected properties of the universe and gives a more complete appreciation of an array of cosmological conundrums we have to explain if we want to understand the overall structure of the universe in which we live.

Past horizons

Let us suppose the universe began expanding at some particular moment in the past, and at that moment a clock began measuring time. Let us also suppose, along with the chaotic cosmologists, that different points in space were endowed with entirely different expansion rates and matter densities at this moment. In order to coordinate these imbalances, there would have to exist ways of transferring energy and material from one place to another. This cannot be done instantaneously; it takes time for such exchanges to occur.

They can occur most quickly when the energy and information is transferred at the speed of light, 3×10^{10} cm s^{-1}, which is about 186,000 miles per sec. No information about the nature of the local space and time properties can be transmitted more rapidly than this cosmic speed limit. This tells us that after the universe has been expanding from its beginning for a time t seconds, there has only been time for signals to propagate a distance equal to the velocity of light multiplied by the time t.

After one second of expansion, which is the latest time at which neutrinos might be able to provide efficient smoothing of an initial chaotic inhomogeneity into regular, isotropic expansion, the maximum distance over which the properties of the universe could become coordinated is only one hundred thousand kilometers, or about the size of the planet Jupiter. By now the expansion of the universe has stretched that coordinated region by a factor 10^{10} but today we observe spatial uniformity in the universe over distances as great as 10^{23} kilometers. Coordinating processes and friction had inadequate time to allow these enormously separated portions of the universe even to know of each other's existence. Why then is it that they have the same average density to within at least one tenth of a percent?

This property of the universe that it is divided up into coherent domains small enough for light to have crossed is called its *horizon structure*. The present size of the cosmological horizon is about 10^{23} kilometers. It encloses and defines what is called the *observable universe*. It is the only portion of the entire universe from which light signals have had time to reach us.

Besides placing a strong constraint on the scope of any smoothing processes envisaged by the chaotic cosmologists, the presence of cosmological horizons makes the observed symmetry of the universe appear even more astonishing. Suppose we represent position in space by a horizontal axis and increasing time by the vertical axis increasing upwards (see Figure 5.1). Light signals move along the broken lines. If we are sitting at the point *O*, we are only causally influenced by the events in space and time that lie inside and on the cone *OCD*. This is called our past light cone (in our figure it is depicted as a triangle because we do not draw in the other dimensions of space).

Now, let us follow the path of some point in the universe as it ages from the initial "big bang." By the time the universe expands to its present age at *O*, observers there can be influenced by the portion *CD* of the initial singularity. The regularity of the microwave radiation can be considered in this setting. Let *AB* correspond to the time when the mi-

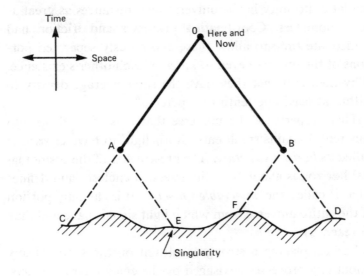

Figure 5.1 *Our Past Light Cone.*

168

crowave photons ceased scattering with matter and began their free propagation towards us carrying with them a picture of the universe's structure at that moment of last scattering when the universe was little more than a million years old.

Our observations of the microwave radiation's similar temperature over different regions of the sky are telling us that points *A* and *B* in space and time had almost identical characteristics—similar temperature, density, and expansion velocities. But if we examine the regions of the initial singularity that determined conditions at *A* and *B*, we see they are in the sections *CE* and *FD* respectively. These are disconnected at the beginning of the universe. There is no means by which the conditions at *CE* can be made to influence those at *FD* or even know that they exist until the universe has existed for long enough to allow light signals from *CE* to reach *FD*. That moment occurs at *O*.

Today, portions of the microwave background radiation separated by an angle of more than fifteen degrees have never been in any form of causal contact during the entire past history of the universe, yet turn out to share the same physical characteristics to within one part in a thousand. It is as if you invited one thousand guests to a banquet with no instructions regarding dress only to find that they all arrive wearing identical dinner jackets. It could be just an accident, but there is likely to be a much more complex reason. Chaotic cosmology envisions a "smoothing" process: all the guests visited each other to coordinate dress before the dinner occurs. The horizon problem insures that some guests live so far apart that they don't have enough time to reach their colleagues in the short interval of time between receiving their invitations and the day of the banquet. Only near neighbors would be able to coordinate their dress. The alternative, quiescent cosmology hypothesis of regular starting conditions would try to persuade us the guests are so similar in temperament and taste (or perhaps that no variety of clothes are available in the shops), that they just independently pick the same dress. A third possibility is that our diners are telepathic—that something in addition to the known laws of physics is at work. Indeed, cosmology encounters the possibility of an epoch when new physical laws may conceivably come into play. During the first 10^{-43} second of the universe's expansion, that period when the entire universe behaves as a quantum gravitational state in a way our current theories cannot describe, signal propagation may not be limited by the velocity of light. Horizons may then disappear, and as yet unknown quantum gravitational processes could coordinate the different parts of the universe.

We can actually squeeze some more insight from this illustration. An unusual suggestion as to why the universe is so regular when surveyed over its largest regions is that the extent of the regularity might be an illusion. Instead of imagining the whole universe of space extending like a flat sheet of paper, suppose we stick two opposing sides of the paper together to form a cylinder (see Figure 5.2). This does not alter the local appearance of space—a tiny region of the curved cylindrical surface still seems flat just as the local portion of the earth's curved surface we are sitting on at this moment seems flat. But, on the whole, a significant change has occurred, namely a change of topology.

A light signal propagating over the flat space keeps on going, but in the cylindrical space it winds around and around indefinitely. In the second case, we think we are seeing out to large distances because the spiralling path of the light rays get longer and longer as the universe ages, but in fact, we are seeing out only to a fixed distance that is the

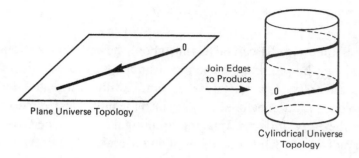

Figure 5.2 *Topologies.*

radius of the cylinder. We are seeing the same images over and over again at different moments in their history as the light rays spiral around the cylinder. The reason our dinner guests look so similarly clad might be because there are really very few of them and the walls of the banqueting hall are mirrors! This type of universe with a peculiar topology is not taken very seriously by cosmologists because it is so hard to test by observation. The best we can do is determine the minimum size of a cylinder producing ghost images by picking a very distant bright object in the sky, like the Coma cluster of galaxies, and search for multiple images of it. This procedure places the minimum size of a cylindrical universe very near to the size of the entire observable universe with simple flat topology. There appear, as yet, to be no other significant observable consequences. Perhaps this conclusion may change in the future as our knowledge of the universe is extended.

Heat from chaos

Something else can be learned about the beginnings of our universe by probing the chaotic cosmology hypothesis in yet another way. Again, let us take a familiar illustration to build up our intuition before plunging into the unfamiliar territory of the early universe. Imagine spinning a bicycle wheel very rapidly and then suddenly applying a brake. The energy of the spinning motion seems to suddenly disappear. But if we were to touch the brake blocks, we would realize that the energy of motion has not been lost completely, merely transformed into heat energy. This is a characteristic of all frictional processes. They degrade ordered forms of motion into disordered heat radiation. The degree of disorder is measured by a parameter called *entropy,* and in all physical processes the entropy must increase. This rule is the second law of thermodynamics.

If we could measure the amount of heat energy in the brake blocks by taking their temperature and estimating their heat capacity, we could gain some idea of how much rotational energy the wheel originally possessed. We could not discover the exact amount because some energy will have been lost in other ways—used up in overcoming air resistance, making noise and so on—but we could certainly discover the *smallest* rotation speed the wheel could have had that is consistent with the amount of heat deposited in the brake blocks.

The microwave background radiation is the heat reservoir of the universe. Any heat radiation generated by the action of friction and dissipation during the universe's early history is degraded into this form of energy. If the universe did begin in a completely chaotic state and, by some unknown means, managed to dissipate these chaotic motions and become regular, the heat entropy generated could not be hidden. The more chaotic the early universe, the more dissipation would be necessary to smooth it into its present state. More heat radiation would be produced in the process, and so the higher the present radiation temperature would be.

A measure of the entropy of the universe is provided by the ratio of the number of photons to baryons which, as we have already seen, is now close to a billion in value. This number limits the amount of chaos that could have been removed from the universe's expansion after the moment when the baryon number was first imprinted upon the universe when only 10^{-35} second old. If gross irregularities in the universe had been smoothed out by frictional processes that obey the laws of thermodynamics and increasing entropy, then they must have completed their work and dissipated all the chaos by this extraordinarily early moment. If they acted later than this to any significant extent, the heat production would have been excessive, perhaps so much so that life would not now be possible in the universe.

Taken together, all these arguments give us a powerful conviction that there is something very remarkable about the large-scale isotropy and homogeneity of the universe; something that is not explicable by the simple physical processes we already understand, something that is intimately connected with the events that initiated the expansion of the universe. This may seem like bad news for anyone hoping to explain the observed properties of the world by general principles and laws of nature rather than by special starting conditions, which are by definition inaccessible to the scientific method. But there might be a silver lining even in this cloud. If the present structure is predominantly determined by an initial configuration and inception, it suggests that we might be able to learn something of events at the earliest imaginable moments by examining the vestigial remnants and cosmic fossils we see around us today in the form of galaxies, radiation fields, and clusters. The chaotic cosmology theory precludes this possibility, but the quiescent picture allows it.

Einstein's universe models

So far in this chapter, we have talked a lot about other possible models for the universe, ones that are chaotic rather than symmetrical, very hot rather than cool with unusual topologies and anisotropies. Where do we learn about such possibilities and what do they look like? Our experience with Einstein's general theory of relativity shows that it updates Isaac Newton's theory of gravitation in a remarkable and complicated fashion. Although Newton's theory is entirely adequate for the type of calculations we might need to make here on earth to build houses, to fire missiles, or hurl footballs, Einstein's theory becomes essential when we deal with systems possessing strong gravitational fields and with objects that move very close to the speed of light. The theory has been tested very carefully by a variety of methods to check its unusual predictions, which differ from those of Newton's theory. These include light-bending by gravity, the detailed orbital motion of the planet Mercury around the Sun, the delay of radio signals from space probes and, in the near future, the precession rate of gyroscopes orbiting the earth. Einstein's theory has been found to be accurate in all of these striking predictions of what are essentially new physical phenomena. These successes, in combination with the appealing elegance and power of Einstein's ideas, have convinced physicists of its excellent approximation to reality. No experimental fact has yet been found to contradict its predictions. When it is superceded, it will be by a theory that must be very, very similar and which includes all its successes.

For these reasons, cosmologists equate possible models for the universe with solutions of Einstein's equations that exhibit expansion in time. There are many such solutions. This is itself an interesting fact since it indicates that something more than Einstein's theory is necessary. The universe is, by definition, unique. It is the only universe we know, and the only universe we can know, so there should exist some extra principle or constraint that proves that all solutions of Einstein's equations except one are irrelevant or impossible. No such sifting principle is known; therefore we must attempt to find solutions that represent an expanding world which most closely resembles our own. The simplest, completely symmetrical model universes are the easiest to find and also provide the best description of the universe in its present state. These isotropic and homogeneous models are those first discovered by Alexander Friedman in 1922. They involve two arbitrary parameters that must be found by measuring the expansion rate and material density in the universe today. It is one of the major goals of observational astronomers to determine the values of these parameters as accurately as possible.

Is the universe rotating?

We see rotation on practically every dimension from the 10^{-13} cm of a proton to the 10^{23} cm of a galaxy. One wonders, with good reason, whether the universe itself could be rotating. Friedman's universe does not rotate but there exist other cosmological models that do rotate. Some rotating models possess other bizarre properties, one of which is the possibility of time travel. Sadly, perhaps, we know that our universe cannot possess a significant level of rotation. More precisely, the maximum allowable rate of rotation is only one ten-thousandth of a single revolution during ten billion years—roughly the time since the "big bang." This rather strong statement is possible because even a slow cosmological rotation has the effect of slightly squashing the universe—just as the earth is slightly wider at the equator than at the poles because of its diurnal rotation. When we scan the microwave background radiation, it reaches us from a great distance where it was last scattered by matter. This happened, on average, about one million years after the "big bang." We regard the matter in the universe at this epoch as effectively constituting a surface that we study by looking at the radiation. This surface of last scattering would be distorted by any global rotation. Observations of the large-scale temperature uniformity of the cosmic background radiation show that no such distortion of the last scattering surface is present.

In 1949, the famous mathematician, Kurt Gödel, found a remarkable solution to Einstein's cosmological equations. Gödel's universe was a nonexpanding one that was rotating. This was not a description of our universe, but it was important nevertheless because of its unusual properties, ones that solutions to Einstein's equations were not suspected to possess. The most unusual of these properties was the possibility of time travel. In Gödel's universe, there is a path through every point in the universe which, if followed, would allow you to travel backwards in time. You could, as the old conundrum has it, kill your own grandmother.

Following Gödel's pioneering work, the extent to which this possibility exists in physically realistic universe models has been examined in some detail. It seems that, unless matter is made to pass through a singularity, or antigravity exists, then one cannot build a time machine from a finite quantity of ordinary material unless another time machine already exists.

Mixmaster universe

If we so choose, we can look for even more unusual candidates as models of the universe's dynamics—applicable perhaps during its earliest stages when we do not know how irregular the expansion might have been. We are confronted, however, by a formidable practical difficulty. Newton's famous theory of gravitation is comparatively simple. It consists of just one differential equation for *one* parameter describing the gravitational force. Einstein's theory which supercedes it, on the other hand, has ten intricately interwoven equations for *ten* gravitational field parameters. Statistics alone should persuade us that it is not going to be an easy matter to find solutions to Einstein's equations! As though this were not bad enough, there is another nasty property of Einstein's equations that provides an almost insuperable barrier to solving them. Einstein's equations are *nonlinear.* This means that the sum of two solutions to Einstein's equations is not another solution. By contrast, Newton's single equation for the behavior of gravity is linear. The sum of two solutions is also a solution. These complexities insure that any solutions we are able to find of Einstein's equations necessarily involve some special symmetry or idealization that either simplifies them and reduces the number of the equations we have to solve or moderates the extent of the nonlinearities.

Cosmologists have no idea yet as to the nature of the general and fully complete solution to Einstein's cosmological equations. The asymmetrical models for the universe that have been discovered all possess some particular symmetries. One large class that is completely understood and classified describes worlds that are the same from place to place but which expand at different rates in different directions. These are the so-called homogeneous but anisotropic universes. The simplest member of this class has played a major role in studying the consequences of cosmological models that expand asymmetrically during their earliest stages prior to the formation of galaxies and stars. It is named after the American mathematician and scientific writer, Edward Kasner, who discovered it during the 1920s. Kasner's cosmological model describes a space that has an ellipsoidal shape at any moment in time; however, the degree of ellipticity changes with time. More remarkably, the Kasner universe expands only along two perpendicular axes; along the third it contracts! The rates of explosion and implosion are such that the overall volume of space expands directly in proportion to time, but the imploding direction means that it evolves towards a state that looks like an ever-flattening, expanding pancake. Of course, the present expansion of the universe looks nothing like this. If it did, we would observe blue shifting rather than red shifting in the spectra of distant galaxies over one third of the celestial sphere and gross differences in the temperature of the background radiation from one direction to another. In fact, we know that the universe could not have been like this any later than one second after the "big bang," otherwise our computations of the amount of helium in the universe would be very different from that observed and predicted by the isotropic Friedman model. Kasner's universe illustrates what the simplest deviant universe might look like within less than a second of the "big bang." What about the most complicated candidate to emerge from Einstein's equations?

The most complicated possible model universe we know about, dubbed the Mixmaster universe for reasons that will soon become clear, by the American cosmologist Charles Misner, is quite extraordinary. Whereas the simple Kasner model has an analogy among the solutions of Newton's simple theory, the Mixmaster universe possesses no such distant cousin. It is entirely an artifact of the extreme nonlinearity of Einstein's theory.

The Mixmaster universe is finite in size. It evolves from an initial "big bang" singularity to a final "big crunch" of a similar character. But if we follow it backwards in time towards its beginning, an extraordinary sequence of events occurs.

At first, it looks like the simple Kasner universe. It collapses in two directions but expands in the third while the overall volume gets steadily smaller. But suddenly, it transforms into a different Kasner model. The expanding axis starts to contract, and while one of the two contracting axes now expands, the other continues to contract. These two axes run through a sequence of these interchanges while the third just decreases steadily. The universe behaves like a contracting ball that shrinks to a thin spindle and then puffs out first in one direction and then in another, changing into a pancake configuration, and so on. Any shape could eventually arise.

But this is by no means the whole story. After these two axes have undergone a series of oscillations from expansion to contraction with the third direction steadily imploding, there arises a further spontaneous exchange of roles. The axis that has not been participating in the oscillations swaps roles with one that has and begins a new sequence of alternations between expansion and contraction while the formerly oscillating axis now steadily implodes. This exchange between periods of steady implosion and rapid oscillation continues sequentially *ad infinitum*. An infinite number of these oscillations will occur before the "big bang" singularity is reached and the amplitude of the oscillations gets steadily bigger and bigger as the overall volume of space gets smaller and smaller. Figure 5.3 displays how the three radii *1, 2,* and *3* of an ellipsoidal ball of material change as one squeezes it to a point in this model of the universe.

The presence of these perpetual sequences of oscillation is strange enough, but it is not the reason for its name as the Mixmaster. This name stems from the analogy with a particular domestic cooking appliance and refers to another unusual property. The Mixmaster universe is a cosmology that might not always have possessed the inhibiting horizon structure we discussed above. If this were the case, it might have enabled the universe to "mix," or become coordinated over great distances at a very early time in its expansion history. Unfortunately, this imaginative idea of Misner's did not work out when examined in detail. Although the Mixmaster model could have enormous extent in one direction and only infinitesimal radii in the other two—good conditions for "mixing," hitting upon this configuration turns out

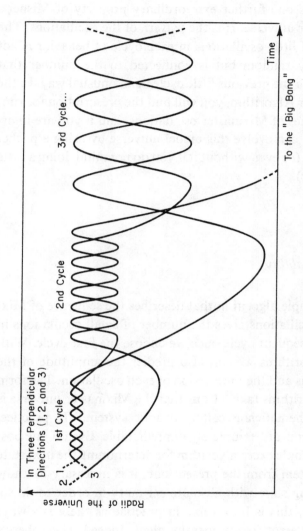

Figure 5.3 *Mixmaster Universe.* The changes in the radii of the Mixmaster universe in three perpendicular directions as it approaches the "big bang." Each of the radii oscillates in size about the others and the model passes through a series of oscillations of increasing amplitude.

to be as improbable as assuming the universe began expanding in completely symmetrical fashion initially.

There is one further extraordinary property of Misner's Mixmaster universe. It is the pattern of the oscillations. The number of little oscillations in each cycle of behavior is not completely random but is connected to the number that occurred in the previous little cycle in an unusual way. In the Mixmaster Algorithm, you will find the prescription for running a series of Mixmaster oscillations, and if you are interested, you can evolve this model universe by using a pocket calculator (or even without it if you do not mind doing a little arithmetic).

Unpredictability

The simple algorithm that describes the sequence of Mixmaster oscillations gives the number of small oscillations in every subsequent cycle once we choose the first cycle. With other algorithms we can also predict the amplitude of the oscillations and the time period of each oscillation. The form these algorithms take tells us that the Mixmaster universe is what mathematicians call a "chaotic system." In practice, such systems are completely unpredictable. Even if one possesses a simple *exact* algorithm for determining the next state of the system from the present one, it is necessary to know the starting state with absolute certainty in order to do so. Of course, this is impossible in practice as there is always some tiny error in any measurement. Indeed, Heisenberg's uncertainty principle insures that it is impossible in principle because the very act of measurement introduces inevitable

MIXMASTER ALGORITHM

Figure 5.3 illustrates that the Mixmaster model runs through a series of oscillatory cycles as the universe collapses down to its initial singularity. A simple algorithm can be determined from Einstein's theory of gravitation that allows one to predict the number of small oscillations in each cycle, as follows:

1. Pick any random number on your desk calculator; as an example, I have

   $$N_1 = 6.0229867$$

   This is the starting condition for the model.

2. Divide the number N_1 into its whole part and decimal part

   $$N_1 = 6 + 0.0229867$$

   Then, in the first cycle, there will be 6 small oscillations.

3. The number N_2 coding the next cycle is found from N_1 by calculating the reciprocal of the fractional part, so

   $$N_2 = 1/0.0229867 = 43.503417$$

   Therefore, the next cycle has 43 oscillations.

4. The third cycle has

   $$N_3 = 1/0.503417 = 1.9864248$$

   It, therefore, has only one oscillation . . . and so on.

You should check that the next 4 cycles have 1, 72, 1 and 5 oscillations respectively. If you had started the model with N_1 just slightly different—say you change the last decimal entry from 7 to 6 and repeat the sequence of calculations—you will find quite a different set of cycle lengths. This illustrates the essential feature of a chaotic system: a very small change in the starting conditions creates a huge difference in the later state of the system. This is one reason why it is so difficult to predict the weather accurately.

uncertainties. What this means for the Mixmaster universe is that if there were even the most infinitesimal of errors— let us say at the tenth decimal place—in our knowledge of the starting conditions, then after only eight cycles of oscillation that initial error would be so amplified that we would be unable to predict anything meaningful about the number of oscillations in subsequent cycles. This is also reminiscent of our stones reaching terminal velocity when falling from the cliff. They quickly lose memory of their starting speed and settle down to a predictable constant speed. The Mixmaster universe also loses memory of its initial state very quickly, but its terminal state is unpredictably chaotic. One can only make statistical estimates of its nature.

A more everyday example might again be helpful. The reader has probably played a cue game like pool or billiards, and so will be only too well aware of the fact that this game, like the Mixmaster universe, is extremely sensitive to its starting state. A catastrophic error results from the slightest miscue. Suppose we forget about the effects of air resistance and friction that stop the ball moving on a real table. If we could hit the cue ball with absolute precision, we could, using Newton's laws of motion, predict the subsequent position and speed of all the balls exactly. This is what the eighteenth-century French mathematician Pierre Laplace had in mind when he claimed that theoretically a "supermind" could predict the whole future course of the universe given the initial conditions. He did not realize, however, that precise knowledge of the starting state is impossible.

Suppose we could know the starting state even as well as quantum theory allows. This would then enable us to reduce our uncertainty about the cue ball's position to an accuracy billions of times smaller than the size of an atomic nucleus. Yet after only about fifteen collisions with other balls, this infinitesimal uncertainty expands to the dimension of the entire table. After that, we know nothing about the subsequent positions of the balls; Newton's laws are quite useless for this purpose. They tell us merely how rapidly our uncertainty is growing. In past times, Newton's laws were often exhibited as the paragon of determinacy; as evidence that the world runs like clockwork in a fashion that is entirely predictable, once we write down the magic equations. In recent years, this confidence has been largely eroded. Many interesting systems, like the Mixmaster universe, are described by exact equations and are not subject to random fluctuations, but become completely unpredictable after a short period of their evolution. Randomness, so long associated with quantum theory, is part of the warp and woof of both classical everyday physics and the earliest moments of the "big bang."

What is time?

Besides teaching some very important lessons about predictability, the Mixmaster universe also tells us something very profound about the nature of time. With the theory of special relativity, Einstein taught us that time is not absolute. The rate at which time passes depends on your motion and the strength of gravity in your vicinity. Rapidly moving clocks move slowly relative to stationary ones, and clocks placed in strong gravity fields are seen to move more slowly than those in weak ones. These unusual effects are now almost commonplace to physicists and are observed. They are not associated with any mechanical malfunction of the clocks. Time really is passing at different rates in different environments.

The twin paradox illustrates the effect. If two twins with identical birthdays are separated, one placed in a weak gravity field while the other in a very strong one, when they are eventually reunited, it would be possible to discover that the twin who lived in the strong gravity field was still in his youth while his brother had reached old age. Of course, the differences in time are minute unless the gravity fields are very strong or the speeds close to that of light, and so these phenomena have no perceptible effect upon our everyday lives. Nevertheless, the rate of flow of time depends on the motion of its measurer. It is relative, not absolute. To the relativist, a day really is as a thousand years.

The only meaningful and unambiguous time is called *proper time*. It is that time kept by a clock moving along with the observer and so not in motion relative to him in any way nor experiencing a different gravity field. When we talk of the age of the universe, we mean the age measured by a hypothetical observer who expands along with the universe from the singularity to the present. Some scientists have raised an interesting question in connection with this definition. They point out that we have no grounds for believing that special relativity applies to the universe as a whole. Also, our observer moving along with the universal expansion is hypothetical in more ways than one. If we follow him backwards in time to the early universe, we will have to replace him or her with an atomic clock when conditions become inhospitable for life. Sooner or later it will be too hot for atoms and nucleons—and everything made of matter.

What do we mean by time when there is nothing left to measure it with? Perhaps there is a universal clock that does not measure proper time. Such a clock would have to be tied to something that is always there, right down to the singularity. The only good candidate seems to be the curvature of space or, equivalently, the density of matter. Everyone in the universe could use this clock. If we believe this curvature clock to measure the real time, then we are led to some very unusual conclusions.

As we approach a space-time singularity, curvature and density probably become infinite, and our curvature clock will measure an infinite amount of "curvature time" to have elapsed during a finite interval of proper time. Suppose we ignore the complications of quantum effects before the Planck moment in the universe and imagine that an intelligent creature exists who can sense curvature time. That is, his subjective or psychological time is experienced at the rate of flow of curvature time. This creature would be able to live forever in a closed universe containing a future singularity! There would be an infinite number of things happening in his future. More extraordinarily, the tables are completely turned in an open universe. Here the observer living by proper time considers himself to have a potentially infinite future, but, according to curvature time, he has only a finite future. His psychological and subjective time will slow down along with the gradually decelerating universe. A concrete example of this type of curvature clock is the Mixmaster universe we have just encountered. If each Mixmaster oscillation of the universe is a tick of curvature time, then a Mixmaster-type singularity must always lie in the infinite past of curvature time because there are an infinite number of Mixmaster oscillations in any finite interval of proper time from the singularity. As yet, there is no way of deciding whether curvature or proper time are preferred in some profound way, but in the meantime, it is conventional to assume special relativity applies to the universe as a whole and that proper time is the meaningful cosmological time.

The Left Hand of Creation

Future

Our ability to predict the future and our knowledge of the
past history of the universe stem from the day when Edwin
Hubble realized that the shifts in the spectral lines of distant
galaxies were related in a systematic manner to their distances. The expansion of the universe was probably the most
extraordinary discovery in the history of science. From Hubble's work, there emerged a model of the universe in which
the past, the present, and the future must be radically different. It is a cosmological picture that suggests a beginning to
space and time, that predicts an early era when the quantum
theory controls even the forces of gravity and provides a
theoretical laboratory in which to study the energetic world
of elementary particle interactions. We have since discovered
that the expansion has many unusual properties. Up to now,
we have talked mostly about its isotropy and uniformity, but
it has another unexpected feature.

In chapter 2, we discussed the notion of escape velocity,
the speed a rocket must attain to escape the gravitational pull
of the earth; more slowly moving projectiles inevitably fall
back to earth. We can picture the expansion of the universe
in an analogous fashion. The initial "big bang" expands the
material in the universe. Hubble first observed that the distance between any two distant objects is steadily increasing.
Yet, this expansion may not always continue. There is an
escape velocity for the expansion of the universe. If the expansion begins with a value that is even infinitesimally in
excess of this escape velocity, the universal expansion will
continue for all future time. If the initial expansion velocity
is less than the critical value for escape, then the gravitational
pull of all the matter in the universe will gradually deceler-

188

ate, and then finally halt and reverse the expansion into recontraction at some time in the future. In our world, this moment lies at least ten billion years in the future. If astronomers still exist, at this time, they will begin to find the red shifts of distant galaxies changing to blue shifts. The way we can tell on which side of the critical dividing line our universe sits is to measure and compare the actual density of galaxies and other material in the cosmos with the velocity of expansion. This is not as easy as it might sound because a large amount of matter might be invisible. Also, many problems make the unambiguous measurement of the recession velocity difficult when the receding objects are very distant. What *is* very clear though is that the expansion rate is very close to the critical divide that separates a destiny of recollapse and infinite compression from one of indefinite expansion, cooling, and rarefaction.

So close is the universe to this watershed that it has so far proven impossible to conclude definitely which of these fates might be in store for us. Indeed, if you scan recent issues of any astronomical research journal, you will find different scientists using different observations and arguments in an attempt to pin down the precise values of the present expansion rate and deceleration. They will often get quite different answers, not necessarily because some are making mistakes in their arithmetic or measurements, but because each type of investigation has its peculiar forms of systematic bias and innate uncertainty. Some methods of analysis always lead to overestimates, some to underestimates. Unfortunately, we do not always know which are which.

The fact that the expansion is, even after ten billion years, irresolvably close to the critical cosmic threshold separating a finite from an infinite future is quite a mystery. It implies something quite extraordinarily improbable about the starting conditions at the "big bang." Imagine trying to set up a model of the expansion of the universe. You would have to be very careful early on so that it does not deviate significantly from the critical escape velocity by the present time. To make the model match what is observed, we have to choose the recession velocity to lie within one part in 10^{28} of the escape velocity at the Planck time, 10^{-43} second, when the expansion emerges from the quantum era. Some very powerful constraint must exist to enforce this degree of fidelity.

Inflation revisited

A clue as to why the expansion should be so finely tuned may ultimately be tied to the discovery that there might arise a grand unification of the strong, weak, and electromagnetic forces during the first 10^{-35} second of the universe's life. From this, there has emerged a radical explanation which has provoked a flurry of interest by both cosmologists and elementary particle physicists. It is called the *inflationary universe* hypothesis. We first encountered some of its consequences in the last chapter. Here is a recapitulation of some of the highlights.

Ordinary substances exist in a variety of states called 'phases'. The most familiar example is the combination of two hydrogen atoms with a single oxygen atom which we call water in its liquid phase, ice in its solid phase, and steam in its vapor phase. If water changes from one phase to another, heat must either be supplied or lost to effect the change. When a liquid changes to the vapor phase, heat is extracted from the surroundings. (For example, your hand will feel cool when ether is suddenly evaporated from it.) Conversely, when a liquid freezes to solid, heat is given out. In a similar fashion, the aggregates of elementary particles like quarks and leptons that populate the very early universe can exist in different phases. When the expansion cools the "big bang" to the temperature at which the strong force becomes distinct from the weak and electromagnetic forces, a phase change is possible in the mixture of quarks and leptons. The phase change will release a large quantity of latent, previously hidden, heat into the universe, and the pressure of this radiation can cause a sudden and dramatic inflation of regions of space. The expansion of the universe will accelerate very rapidly and expand out to a dimension enormously in excess of what would have been reached at these early times if a phase change had not occurred.

If this inflationary process occurs, it might also explain the present close proximity of the universe to the critical escape velocity. Suppose the universe began with an expansion far below the critical escape velocity, but after about 10^{-35} second, the complete symmetry between the strong and electroweak interactions broke down, accompanied by a phase change and concomitant heat release. The effect of the dramatic acceleration of the universal expansion during a period of inflation is to increase the expansion rate until it is very very close to the critical rate dividing universes that expand forever from those that will recollapse in the future. The longer the period of inflationary acceleration lasts, the closer the universe would get.

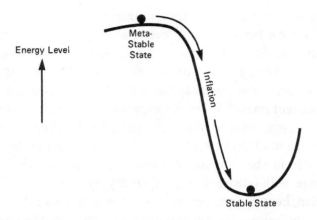

Figure 5.4   *A Phase Transition in the Universe.* At first the universe resides in a meta-stable state when all the interactions are unified. As it cools, the symmetries between different forces are broken and it moves to a stable state of lower energy. Inflation occurs during the slow transit between the two states.

Inflated horizons

While the inflation occurs during the transition phase, the regions of the universe that can communicate with one another grow dramatically larger. This means that frictional processes can operate to smooth out preexisting irregularities. If there is an excess of radiation in one region, it will diffuse away. Regions of dimension up to the distance traversed by light at the end of the inflation can be smoothed out. This scale may exceed the size of that portion of the universe we can now see. Small fluctuations should still remain. We saw in chapter 4 that the surviving hetereogeneities may be strong enough to provide the seeds from which galaxies later formed. In this way, it is possible that inflation can account for the apparent paradox of the universe, why it is almost, but not quite uniform. [13]

Other universes

If the early pattern of expansion were sufficiently irregular, some regions of the universe could have already collapsed. If one of these collapsed regions is smaller than the horizon size, a black hole would form. If it were larger than the cosmological horizon size at the moment of collapse, a separate closed universe would result. Events within the closed universe would have no influence upon the outside universe and would remain forever unobservable from outside. We could neighbor one, or indeed many, of these separate closed universes. Such worlds would be collapsing by now, since they must have formed long ago and possess a density much greater than our own universe. The domain of space-time occupied by these universes is inaccessible to us. If we were to look towards them, we would think we were observing events at the beginning of our own universe. Actually, the inflationary scenario allows an infinity of such universes: gigantic bubbles, each much larger than our present observable universe. Each should be very smooth but every one could be completely different.

Einstein's greatest blunder

In 1917, Einstein introduced the first modern cosmological model—a nonexpanding, static universe. He realized that the attractive force of gravity would render it unstable against collapse. To balance gravity he chose, not universal expansion energy, but a new cosmological force of repulsion.

He augmented his original cosmological equations with a term known as the cosmological constant. The constant was very small in magnitude and only had a noticeable counter effect against gravity over distances of billions of light years. Einstein introduced this new force of nature because he erroneously believed that the universe must be static. In 1922, Friedman noticed Einstein's mistake. Einstein had divided both sides of a key equation by an expression that could be zero. If it were zero, Einstein's proof would break down. By noticing this possibility, Friedman was led to display the existence of both expanding and contracting universe models in Einstein's theory. The introduction of the cosmological constant was not necessary after all, and it was that idea that in 1931 Einstein called "the biggest blunder of my life." Once introduced, like the opening of Pandora's box, the possibility of cosmic repulsion complicated cosmologists' lives. Its most enthusiastic supporter was Arthur Eddington, a British astronomer, who advocated a model that expanded from a static Einstein state. Another supporter was the Abbé Georges Lemaître, the Belgian cosmologist who preferred a universe that began as a singular "primeval atom" and expanded into a temporary static phase before resuming expansion. Both models allowed galaxies to form very easily during the coasting phase because there is no expansion to resist the condensation of matter into lumps.

In recent years, a universe model incorporating the cosmological constant that was first proposed by the Dutch astronomer Willem de Sitter has been revived. Originally proposed as an empty expanding counterpart to Einstein's static world, it provides an exact description of the inflationary expansion triggered by the phase change at the epoch of cosmological grand unification. The vacuum energy released during the phase transition temporarily introduces a huge cosmological constant into the universe. One can show that the vacuum energy must be formally equivalent to Einstein's cosmological repulsion force; only in this case does it look the same to all possible observers in the universe no matter what their motion. Once inflation is over, however, after some 10^{-30} second, the vacuum energy has vanished, and the cosmological constant is zero, or very close to zero. It is curious to note that a consequence of this inflationary phase of the universe's early life is that Einstein's "greatest blunder" may yet prove the key to understanding the evolution of the universe. [14]

Magnetic monopoles

Grand unification has a further, potentially disastrous implication for cosmology that inflation may resolve. This arises from the creation of magnetic monopoles. Electricity and magnetism are everyday manifestations of the same basic force of nature—the electromagnetic interaction. In many of their intertwined properties, electricity and magnetism exhibit a beautifully symmetrical relationship, but in one dramatic respect, this symmetry seems to be flawed. Although we see examples of point *electric* charges in the world —the electron, for instance—we never see any *magnetic* charges, or isolated magnetic poles. Ordinary bar magnets possess a north and a south magnetic pole. Whenever we attempt to isolate one of them by cutting the magnet in half, we do nothing more than create two magnets, each possessing North and South poles. Ordinary magnets with two magnetic poles are called *dipoles* and the shape of a dipolar magnetic field can easily be seen by placing a small bar magnet under a sheet of paper sprinkled with iron fillings. If the bar magnet stands on one pole immediately beneath the paper, the new pattern made by the filings is an example of the field exerted by a single magnetic pole—a *magnetic monopole.*

There is no mystery why magnets beget more magnets when cut in half. Magnetism arises from the flow of tiny loops of electric current in atoms, these are generated by the motion of the electrons around the nucleus. Each atom has, in fact, a mini-dipolar character. The magnet's field is just a reflection of this.

Dirac's monopole

In 1931, Paul Dirac, an English mathematical physicist asked himself two questions: Can magnetic monopoles exist, and why does all electric charge come invariably in whole numbers of electron charge units? He showed that the answers were beautifully linked. If magnetic monopoles exist, it would explain why all the electric charges we see come in discrete multiples of the unit electron charge. (The existence of quarks does not alter the conclusion; only now electric charges come in units of one third of the electron charge.) Ever since Dirac's suggestion, there have been experimental searches for magnetic monopoles. Occasionally, someone claims to have found them; only to see them later dissolve as other, more conventional, particles are shown to explain the experimental evidence. The continued lack of success led to a waning of interest in monopoles by the early 1970s, tempered by the nagging worry that if they do not exist, there must exist some veto or law of physics that theoreticians had completely missed.

The GUT monopole

Then in 1975, Alexander Polyakov of Moscow and Gerhard t'Hooft of Utrecht showed independently that magnetic monopoles always exist in grand unified theories of elementary particles. The monopole mass is the equivalent of a little more than ten times the energy scale of grand unification, about 10^{16} GeV. This would certainly explain why they had never been seen: 10^{16} GeV is about 10^{-8} gram, the mass of a bacterium and extraordinarily large for an elementary particle. This ensures that monopoles are very difficult to produce. The encounter between one of these superheavy monopoles and an ordinary atom would be like a steamroller hitting a pea. A monopole that hit one side of the earth could just plough right through and emerge unscathed out the other side. This makes monopoles very difficult to catch.

What does a superheavy monopole look like? It is not a point-like particle as we imagine quarks and leptons to be. Rather, it has a nested internal structure like a series of Chinese boxes. Most of its 10^{-8} gram resides in a tiny core just 10^{-28} centimeter in diameter. Inside that core, energies are high enough for grand unification to exist, and all interactions possess the same strength. Around the boundary of this inner core, there is a shell where X bosons are numerous. Beyond this, there is a sparse periphery fading away beyond about 10^{-16} centimeter where a thin shell of W and Z bosons is predicted. This peculiar structure means that monopoles can affect the stability of other matter in spectacular ways (see Figure 5.5). In chapter 3, we saw how the occasional production of an X boson by a random fluctuation enables a quark to transform into a lepton and induce a proton to decay, on the average, once every 10^{31} years. If a proton runs

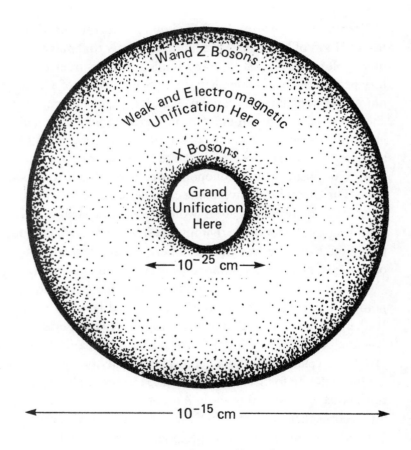

Figure 5.5 *Magnetic Monopole.*

into a magnetic monopole and penetrates its outer shell sufficiently to see the ring of X bosons, its chance encounter with so many X bosons will induce the proton to decay far more rapidly than usual. So monopoles act as a sort of catalyst for proton decay. By calculating the details of how this catalysis occurs, we may be able to detect the tell-tale presence of monopoles in the proton decay events some experimenters have begun to witness in subterranean experiments. An amusing suggestion has been made that the capture of a monopole might solve some of the energy problems of the future. By showing our "tame" monopole some protons, it would transform them into controllable energy with no dangerous waste products.

An observation?

There has been a dramatic new development recently. Blas Cabrera of Stanford University has claimed that he detected a magnetic monopole. He used a superconducting quantum interferometer (SQUID). This is a loop of niobium, just five centimeters in diameter, cooled down close to absolute zero where electrical resistance becomes virtually nonexistent. This loop is the monopole detector and must be shielded from all outside influence, especially the earth's own magnetic field. If a monopole passed through the detector, it would produce a definite signal. Cabrera reported that at 1:20 A.M. on St. Valentine's Day, February 14, 1982, his apparatus recorded a monopole signal after 150 days of searching. If this detection really was of a monopole and not some subtle external noise, then the significance of its discovery is immense. For, if a measurement of a monopole's mass could be made by watching how its path arcs as it moves through a strong magnetic field, this would tell us the energy of grand unification directly. [15]

These monopoles cannot have been created very recently since there are no sites in the present universe that are hot enough. They must be fossils surviving from the first 10^{-30} second of the "big bang." But when the first calculations of their vestigial abundance were made, they created a fantastic contradiction between grand unified theories and astronomy. So many monopoles are produced in the "big bang," (roughly one for every proton we now see) and so inefficiently do they annihilate with their antimonopoles, that we should today have a universe containing 10^{11} times more mass in monopoles than in galaxies. This is impossible. Such a universe would be so dense that it would have collapsed to a second singularity and ended after a mere ten thousand years of expansion! That is not the universe we live in. Something must be missing from our picture of the very early universe.

Inflation removes monopoles

The inflationary universe offers a neat resolution to this contradiction. First, how do monopoles form? Random fields of energy arise near the "big bang" and can become erratically knotted and contorted. Subsequently, they can only become correlated and aligned over distances within the range of light signals, since only such smoothing motions occur at light speed. At the boundaries of these regions of communication, there will arise a mismatch in the orientation of the energy fields, and knots of energy will arise. It is these knots of concentrated energy which eventually materialize into the surviving magnetic monopoles. At the time of grand unification between the fundamental forces in the early universe, the sizes of regions linked by light signals are so minute (less than 10^{-25} centimeter across, in fact) that there a huge number of mismatches and knots will arise. This is why such a picture predicts an inordinately large number of monopoles. Cosmic inflation comes to the rescue because it creates a very rapid, temporary increase in the size of regions within which communication is possible. The mismatches are all ironed out before 10^{-35} second, the time by which matter cools to the stage where the monopoles would have materialized. Virtually none will appear because only a tiny number of persistent knots and wrinkles will remain.

The present density of monopoles can easily become acceptably small if inflation takes place, although it is not possible to predict its definite value because it depends on facts about the monopoles we do not yet know precisely, like their mass—or even whether they really exist. Unfortunately, if Cabrera is really seeing magnetic monopoles, we are faced with a very awkward problem. We do not need to know the number of monopoles surviving the period of cosmic inflation in order to say how many could exist today.

The Milky Way possesses a small magnetic field of about a microgauss. If magnetic monopoles are moving around in our galaxy, they will be strongly accelerated by its magnetic field, just like metal objects dropped between the poles of a horseshoe magnet fly to one pole or another. Gradually, this process will drain away the energy of the galaxy's magnetic field, and eventually it will completely disappear, used up in moving the monopoles. The fact that this has not happened and that our galaxy still possesses a measurable magnetic field means that there simply cannot be many monopoles for the field to accelerate around the galaxy. The largest flux of monopoles compatible with the existing magnetic field of the Milky Way is sometimes called the Parker limit after Eugene Parker of Chicago University.

This is where the problem starts. Cabrera's single detection over 150 days, if typical of the monopole abundance in space, corresponds to a flux one million times bigger than Parker's limit. Something is wrong. The magnetic field of the galaxy should not be there! Perhaps it is the experiment that is wrong; statistics from one event are misleading. Perhaps the galaxy has an as yet unknown way of regenerating its decaying magnetism. Or perhaps monopoles can arrange themselves to evade Parker's limit. For example, if monopole orbits focus their trajectories in directions parallel to the magnetic field, there can be a greatly reduced rate of magnetic field destruction. Or conceivably, our sun could have attracted an anomalously large number of monopoles into its vicinity by the force of its gravity, and then our local flux might not be typical of the average monopole population in the galaxy any more than the density of bees around a honey pot is representative of their density in the garden.

Unusual creations

Monopoles are not the only esoteric structures that can be created during the moments of transition from a grand unified universe to a world of weak, strong, and electromagnetic forces soon after the "big bang." One and two dimensional analogues of monopoles can also exist. These random "strings" and "walls" of energy, as they are called, are like infinitely drawn out or flattened monopoles. Like monopoles, they can easily persist to the present if produced in large numbers during the grand unification epoch. But, unlike monopoles, they are not very massive, and we do not seem to face a catastrophic overabundance of them—even without invoking the smoothing influence of the inflationary universe. However, even the moderate number we expect, could create very unusual effects in the observable universe. A large "wall" of energy stretched across a significant fraction of the universe would, by its gravitational pull, slow down part of the cosmic expansion and lead to such a strong difference in the temperature of the 3K microwave radiation in that direction that its presence would easily have been detected. "Walls" are definitely not there.

"Strings," however, are a different matter. Their gravitational effects are even more moderate. They could emerge from the "big bang" as a huge tangled network of thin tubes of energy that are gradually stretched and unraveled by the expansion of the universe. What would these strings of energy look like to the intergalactic traveler? A string born at the grand unification epoch would resemble a thin tangled tube of radiation about 10^{-27} centimeter wide, stretching 15 billion light years across the entire observable universe. It would have the mass of more than ten million galaxies. Such strings do not lead to any observable distortions of the cosmic background radiation. Their gravitational effects are quite modest. Strings might still reveal themselves since some material would tend to be gravitationally focused around them. It has even been speculated that the patterns and tesselations they weave upon the universe could provide an explanation for the filamentary distribution of the galaxies in space.

Man and the cosmos

The ideas we have been delving into in this chapter seem very remote from the universe we see around us today and the probability of our own existence on a ball of matter near a star called the Sun at the edge of the Milky Way. Many a philosopher has argued against the ultimate significance of life in the universe by pointing out how little life there is compared with the vastness of empty space and the multitude of distant galaxies. At first, one might think that the "big bang" theory could only reinforce such a materialistic view of the universe, but this may be too simplistic a judgment. The observable universe contains some one hundred billion galaxies each of which contains at least one hundred billion stars, and unknown numbers of lesser, dwarf galaxies which could collectively contain as many stars again as the bright galaxies. Could the idea that life on earth was in any way special ever be defended? If the universe were "designed" with us in mind, why bother with all those billions of other completely superfluous galaxies? Why not economize and scale down operations? After all, a universe containing a single galaxy has scope for more than a hundred billion solar systems.

Surprisingly, this argument that the size of the universe mitigates against the importance of life is fatally flawed. Hubble's discovery that the universe is expanding means that the size of the observable universe is inextricably bound up with its age. The reason the universe is so big is that it is so old. A small universe would have to be a younger universe and a universe containing the mass in one galaxy alone would be less than a year in age! We have learned that the complex phenomenon we call life must be built upon elements of nature that are more complex than hydrogen and helium gases. Most biochemists believe that the element carbon, on which our own organic chemistry is based, is the only viable basis. In order to create the building blocks of life, carbon, nitrogen, oxygen, phosphorus, and silicon, the simple products of the "big bang" must be cooked at high temperature. The furnaces that nature has supplied are the interiors of stars. There, the hydrogen and helium from the "big bang" is slowly cooked into the heavier elements of life. When stars die, the resulting explosion, seen as a supernova, can disperse these elements through space, and they become incorporated into planets, asteroids, and other forms of interstellar debris. Stellar alchemy takes a long time, over a billion years. Hence, we realize that for there to be time to construct the constituents of living beings, the universe must be more than a billion years old and consequently more than a billion light years in size. The biologist might also tell us that times of this magnitude are necessary to evolve advanced life forms from simple prebiotic molecules. So we should not be surprised to find our universe is so large. No astronomer could exist in one that was significantly smaller.

This style of argument places the controversy over the existence of extraterrestrial life in an intriguing light. Many astronomers like to argue that life must exist elsewhere because there exist so many sites where it could do so. The overriding opinion of biologists, on the other hand, is that the probability is essentially nil because the number of evolutionary pathways leading to biological dead-ends is at least as large as the number of sites. Our argument destroys this potential paradise for extraterrestrials. The universe would have to be just as large as it is to support even one lonely outpost of life. The global structure of the universe is unavoidably bound up with some of the smallest details of life on earth.

This unusual perspective on the universe focuses on an examination of how its actual large-scale properties are related to those that are essential for the evolution and maintenance of living beings. It has been called the "anthropic principle" and can be brought to bear on a number of the cosmic conundrums we have described in this chapter. If the universe were not now expanding very close to the critical state with just enough speed to expand forever, then the possibility of evolving observers in the universe would be enormously reduced and perhaps removed altogether. If the initial speed of the "big bang" had only been tuned to one part in 10^{29} rather than one part in 10^{30}, the expansion would either have reversed and recollapsed before any stars and galaxies and life evolved or would be proceeding so rapidly that galaxies and stars would be unable to form. Worlds expanding much faster than the critical rate would almost certainly be devoid of stars and galaxies and hence the building blocks of which living beings are made.

The anthropic principle has a simple message: the universe has many unusual properties that *a priori* appear extremely unlikely if we consider the whole gamut of possible universes. But if we ask which of those hypothetical, and seemingly more probable universes we could exist in, we invariably discover that the answer is very few indeed. Our astonishment at some of the striking properties of the universe must be tempered by the realization that many of them are necessary prerequisites for the existence of intelligent observers. The simultaneous existence of a whole range of special coincidences of nature, both in the properties of the universal expansion and in the precise values of the fundamental constants that control the strength of the forces of nature, has tempted some to indulge in further metaphysical explanations. Could simultaneous coincidences that allow our existence be telling us that life must in some way be *necessary* in order to make it meaningful?

6

Conclusions and Conundrums

THE aim of modern cosmology is nothing less than a complete reconstruction of the past history of the universe. We have tried to give an account of a story that begins with the creation of space and time and the uncertain events of the elementary particle world and ends with the condensation of galaxies, stars, and planets fifteen billion years later.

Our theories of how the universe unfolded are never proved true. They serve only as guides to their successors. What sort of question will a newer and better theory of the universe have to come to terms with? What weaknesses in our present ideas will any new theory have to repair in order to prove itself? There is one obvious conundrum that our journey from the "big bang" has left unresolved, or perhaps unresolvable; namely, what preceded the event we call the "big bang"?

If the "big bang" really was a universal space-time singularity of the sort we described in chapter two, then it delineates an absolute edge to the entire universe of space and time, and the answer to our question is simple: nothing. Before the singularity, neither space, nor time, nor matter, nor motion existed. Only by hypothesizing an extension to the laws of physics that we know already can the universe be followed backwards in time beyond the singularity, for instance through an eternal sequence of oscillating cosmic cycles each rising, Phoenix-like, from the ashes of its predecessor. If the "big bang" singularity could be replaced by something more moderate, such as an as yet unknown manifestation of a quantum theory of gravity, then this program of "retrodiction" (as opposed to prediction) might be possible. Then, prior to the event we think of as the "big bang," the universe would have been in a state of overall collapse eventually rebounding into the state of expansion we now see. During the next "big squeeze," at the end of our own cycle, all the matter once in stars and galaxies would be pulverized to its ultimate subnuclear constituents, ready to be resuscitated by unknown processes for the next cycle of expansion.

Our answer to the request for information about the universe before the "big bang" should strike the reader as necessarily a little evasive. We are limited to answering questions about the universe only in terms of things we now know and understand, but yet we are still far from understanding all of physics. We might even wonder whether a complete knowledge of physics is sufficient to understand and explain what the universe is. Conversely, we may inquire whether the observable universe contains enough evidence for us to uncover the underlying laws and principles that govern its existence.

Throughout our attempts to reconstruct the evolution of the universe, we have assumed a uniformity in the laws of nature. We have assumed that the laws, which have jurisdiction over events here and now, also rule the there and then. But for all we know, what we call the laws of physics might have been subtly different in the past, or even be different elsewhere today. This sounds like a straightforward idea, and people often consider it, but its meaning is unexpectedly elusive. For if the "laws" of physics as we imagine them today are found to have altered when we look again next week, all it can mean is that we were wrong in our original choice of invariant laws. If our supposed "laws" are found to change, then the rules governing their change can, in principle, always be found, unless the changes are injected at random. The regularity of nature argues against this latter caprice. What we call the laws of nature are simply compact codifications of events that we see, or expect to see, happening. The ultimate and correct laws of nature we are seeking through science are a collection of rules imposed upon the contents of the universe like the rules governing a chess game. Let us watch a game in progress. Even if we have never seen a chess game, we can begin to piece together the rules by observing the way moves are habitually made. We would soon get to know that bishops always move diagonally. We would suggest a "law" that rooks never move diagonally and can never exchange places with other rooks of the same color. Everytime a player put his finger on a piece, we could make a verifiable prediction of the range of moves open to him. Time after time we might be proved correct, and our confidence would grow to a point where we would think our knowledge of the rules was complete. But suddenly, something we had never seen before might occur—castling. We would have to enlarge our conception of the rules, and maybe even throw some of the "laws" away. We would not as a result of this experience, though, say that the rules of chess were changing, but only that our perception of the rules had been incomplete. Or we might not have been watching the game carefully enough at times, and so a move might have been incorrectly recorded. We could then create a spurious law temporarily until we found it to be consistently "experimentally" refuted by later moves.

A change in the laws of physics from place to place in the universe would be akin to playing chess on a board where the rules governing allowed moves depended not only on the piece being moved but also on its position on the board. If you observed this unusual form of chess for a period and consistently failed to predict moves under the assumption that the rules were independent of position, then you would grow more and more certain that your simple assumption was incorrect. This, in effect, is what astronomers would need to do. They have been able to construct good explanations for distant phenomena using the same rules that apply successfully on earth. As yet, there is no evidence to make one suspect any spatial variation in the rules governing astronomical events that would indicate, for example, there are places where energy conservation would not be seen to hold or where the electron is heavier than we measure here and now. This type of variation is especially hard to imagine because junctions would have to exist in space between portions of the universe under different jurisdictions—a sort of cosmic "Berlin Wall." This type of schizophrenia goes against our strong belief in the unity of the universe, but who knows, one day the universe just may "castle"!

One of our goals has been to deduce and explain as much

as possible about the universe by appealing to as few special assumptions about its beginning as possible. Instead of being satisfied with a statement that galaxies exist because the universe came into being with ready-made galaxies, we try to begin by imagining, or demonstrating, that tiny random disturbances or irregularities existed in the initial make-up of the cosmos. Our aim is to show that objects resembling galaxies will then inevitably develop. In the same economical spirit, cosmologists have tried to show that the peculiar isotropy of the cosmic microwave radiation and the favoritism for matter rather than antimatter are inevitable consequences of a few compelling and simple principles rather than artificially contrived properties of creation itself. We can always produce a model of our universe by assuming that its unusual properties were part of its initial make-up, just as our old friend Gosse imagined the Creator making preaged fossils in very young rocks. However, this procedure fails to *explain* phenomena; it merely describes them. It tells us only that things are as they are because they were as they were. Our aim is to begin with simple and, in some sense, minimal assumptions and then to show that the laws of nature alone inevitably give rise to a universe rather like the one we see. We have seen how, in some respects, this goal is achievable, but in many other ways the task is far from completion. The end of such a quest might be made more ambitious still. We could imagine trying to dispense with the laws of nature as well and then appeal to principles of logic and statistics alone to create the ordered world about us. Indeed, one might even imagine that there may not be any laws of nature at all!

Conclusions and Conundrums

Are there any laws of physics?

Human beings have a habit of perceiving in nature more laws and symmetries than really exist there. This is an understandable tendency because it is the business of science to organize the experimental knowledge of the world, and usually excessive organization is better than a dearth of it. During the last twenty years, however, many quantities we traditionally believed to be inviolable have turned out to be changeable. These include baryon and lepton number, parity, and particle-antiparticle symmetry. All have transpired to be only "almost" conserved in natural processes. The neutrino was always believed to be massless, but now there is very tentative experimental evidence and theoretical arguments to suggest that it does, after all, possess a tiny positive mass. Likewise, the long-held belief in the absolute stability of the proton now appears to have no basis other than blind faith. In view of this general trend from more laws to less, it is just possible that complete anarchy may be the only real law of nature. People have even debated that the presence of symmetry in Nature is an illusion, that the rules, governing which symmetries nature displays, may have a purely random origin. Some preliminary investigations suggest that even if the choice is random among all the allowable ways nature could behave, orderly physics can still result with all the appearances of symmetry. This is possible because we only make observations of the world at energies far below the Planck temperature of 10^{32} degrees Kelvin. A form of natural selection may take place in which, as the temperature of the universe falls, fewer and fewer of the entire gamut of possibilities have a significant effect on the behavior of elementary particles. Conversely, as one approaches the Planck energy at the beginning of the universe, things become chaotic and unpredictable. If this is so, our low energy world may be necessary for physics as well as physicists!

If you went out into the street and gathered information for a long period of time, say, on the height of everyone passing by, you would inevitably find that a graph of the number of persons versus height would result in a characteristic bell-shape called the "normal," or Gaussian, distribution by statisticians. It is ubiquitous in nature. This curve is characteristic of the frequency of occurrence of all truly random processes whatever their specific origin. The resulting curves differ only in their width and the mean on which they are centered. A similar universality might conceivably be responsible for the apparent regularities of our low energy world. So, there are some real uncertainties, not only in our knowledge of the laws of nature and their uniformity, but even in the very concept of what we mean by laws of nature.

Any reader who has read this far will have begun to realize that cosmology is not only about stars, galaxies, and the unending vistas of space. The detailed nature of galaxies, the material content of the universe, the events that occurred in the earliest instants are all bound up with the behavior of the most elementary particles. The very largest structures of nature appear to be inextricably linked to the very smallest. Because of this symbiotic relationship between the sciences of the macro-and micro-worlds cosmologists and elementary particle physicists have joined forces in their quest to elucidate the earliest moments. Then, the cosmos resembled a fantastic experiment in high energy physics. This "new cosmology" contributes its own ideas and explanations of what we see. It also carries its own limitations. We need to understand elementary particles completely if we are to decode the mysteries of cosmic evolution. Many of these particles are not visible in the usual sense of the word. Can we know how many elementary particles to expect? What guide do we have to predict their unusual properties and behavior? Until the 1970s, many different ideas had been unsuccessfully pursued in answer to these problems, but now physicists have found a key that seems to unlock the closely guarded secrets of the elementary particle world.

Why are there elementary particles?

Physicists have learned from experiments that the world of elementary particles is constructed along highly symmetrical lines. It is astonishing that many of the mathematical patterns catalogued by mathematicians in past centuries have subsequently proven to be the very same ones that nature employs to orchestrate the number and the nature of its most elementary objects. The present mathematical description of each of the forces of nature is based upon a particular pattern that is maintained when a force affects a collection of particles. Such *gauge theories,* as they are called, are models for the behavior of matter in which observable events are left unchanged by the application of a mathematical symmetry operation. This invariance comes in two forms: the first, termed *global gauge invariance,* demands that the consequences of the theory are independent of changing in the same way the position in space of everything. For example, a global transformation on an apple would just shift all its constituents sideways in the same way, and so leave the overall form of the apple unchanged. Global gauge invariance of physical laws after translation in space or time, or after rotation in space, is the reason for the great conservation laws of momentum, energy, and angular momentum.

From 1967, physicists began to believe that theories ought to be even further constrained by the dictates of symmetry. After all, how does each bit of the apple know how the other pieces are going to move? It might be an awfully big apple, and sending information across it will take a finite amount of time. To avoid the requirement of acausal signaling to synchronize the movements of independent parts requires that the consequences of a theory be unchanged even when every point is given a *different,* independent transformation. This much stronger stipulation is called *local gauge invariance.* All currently successful theories of physics possess this property.

If we think back to our apple, we can appreciate that the only way for the apple to remain intact under a local transformation, instead of rushing off in all directions, is if forces exist that constrain how the pieces can move. In this way local gauge invariance can be used to demand that the observed forces of nature exist. Better still, it turns out that it can also demand that the individual particles exist and that they possess precise characteristics. For example, the local invariance of a theory of electricity and magnetism requires not only that the photons of light exist, but also that they have zero mass and interact with other electrically charged particles in a particular way, just as is observed. The idea of gauge invariance has been the key to unlocking the secrets of the micro-world. The assumption that a particular symmetry is preserved in nature tells us about specific natural laws and about some of nature's constituents and their unique properties.

As we look further and further backwards in time towards the beginning of the universe, we encounter physical conditions of ever-increasing extremity. Composite structures, even atoms and their nuclei, all are eventually dismembered there. Our greatest problem then is to find what the pieces are. What are the ultimate building blocks in the elementary particle world? Until we know the answer, our theories of the first moments of the universe can only be a part of the truth at best.

What are the ultimate elementary particles?

With the construction of very powerful particle accelerators, it has been possible to see what happens when particles collide with each other at energies high enough to shake any internal constituents they might possess. These experiments have confirmed the expectation of gauge theories, that protons and neutrons, along with all the other hadrons, are not truly elementary particles at all but contain subcomponents. When two protons collide, they scatter each other away very strongly, indicating the presence of three hard internal scattering centers. However, when two electrons collide they scatter less vigorously, suggesting that they contain no underlying structure. Experiments like this have established a provisional belief that the internal constituents of the proton —the fractionally charged quarks—along with the lepton family (electrons, muons, tauons, and their associated neutrinos) are indivisible, elementary, point-like objects. All these particles behave, to the extent we can measure them, like points rather than extended structures when participating in high energy scattering experiments. If they do possess a finite size, it must be less that 10^{-15} centimeter, about one

217

hundred times smaller than an atomic nucleus. Cosmology actually leads one to take the idea of infinitesimal, point-like particles seriously. If the true building blocks of nature were the size, say, of protons, 10^{-13} centimeter, then when the Universe was less than 10^{-23} second old, there would have been grave problems. Different portions of these elementary particles would have been causally disconnected. Insufficient time would have elapsed since the "big bang" for light to travel from one side of the particle to the other. It is not clear how such an uncorrelated object could exist, let alone be called an elementary particle.

Even though there is so far no experimental evidence for it, physicists would not be surprised to discover ultimately that quarks and leptons also have internal constituents. One reason for suspecting this is that there are simply so many different quarks and leptons. What began as an exclusive club has exploded in membership to eighteen different types of quark, three types of leptons, along with three types of neutrino—not to mention all their antiparticles. The newest idea, supersymmetric theories, doubles this number. This demographic explosion looks very suspicious. Surely, the ultimate building blocks of the universe are more economic and elegant than this?

The Israeli physicist Haim Harrari has suggested a scheme in which there are just two elementary particles of nature, called *rishons*. He speculates that there is a T rishon that has electric charge of 1/3 and a V rishon that has a charge of 0. All the known particles can be built simply from strings of T's and V's along with their antiparticles, $\overline{\text{T}}$'s and $\overline{\text{V}}$'s. For example, the positron is TTT, the electron neutrino is VVV. Each quark is comprised of the rishons in an order that fixes the color quantum label, and so protons and neutrons are arrays of nine rishons. All known reactions between elementary particles can be transcribed into this rishon language, but this alone does not mean that it is a theory of physics. As yet, no experimental test of the rishon idea has been suggested, and there is no known gauge invariance that demands the existence of the T and the V particles.

Perhaps there are no elementary particles at all, and everything is infinitely divisible, composed in some way of everything else. A theory of this sort, called a *bootstrap* theory, was very popular in the late 1960s before evidence for the existence of quarks was found. The original bootstrap theory had the unusual property of predicting that the maximum attainable temperature is about 10^{13} degrees Kelvin. When energy is added to a collection of particles at this temperature, instead of heating them further, lots of new particles were imagined to appear and the temperature of each did not rise. We know from experiment that this cannot happen at the moderate (by elementary particle standards) temperature predicted by these early theories, but perhaps something similar could occur with an ultimate temperature equal to the Planck value, 10^{32} degrees Kelvin?

The particles we have already seen fit together in symmetrical patterns that predict definite phenomena which can be tested by observation. Our discoveries have shown us that, instead of being an immensely complicated place, the "big bang" was a relatively simple one. The behavior of elementary particles depends strongly on their temperature. As the temperature rises, hidden symmetries appear, diverse interactions become identical, and all dissimilarities steadily disappear. A grand unification of all particles and forces is anticipated. These grand unified theories of the strong, electromagnetic, and weak forces enable us to partially reconstruct events during the first 10^{-35} second of cosmic history, and lead to testable predictions in the realm of particle physics. These theories are not yet complete because they do not contain all the possible symmetries. Neither do they combine the force of gravity with the other fundamental forces, nor do they yet place all types of particles on the same footing. Until such a superunified theory emerges, there will remain doubts as to how greatly our current conclusions would need to be revised by that next step forwards.

What are supertheories?

Supersymmetry, and its "big brother" *supergravity*, are theories born of the continual desire of particle physicists to unify the forces and elements of the universe.[16] There are two distinct types of particles in nature: bosons whose spin is counted in whole units and fermions, whose spin is in half units. Supersymmetry is an idea to create a symmetry between these divided populations. It does this by proposing that there really exists a further collection of, as yet unseen, elementary particles that restores the balance. All the bosons, the photons, gluons, W's, Z's, and gravitons possess fermionic partners, the so-called photinos, gluinos, Winos, Zinos, and gravitinos. Likewise, to even things up completely, the known fermions—the quarks and leptons—have supersymmetric relatives that are bosons, called squarks and sleptons. None of these new particles have been seen. Recently, detailed cosmological consequences of these hypothetical superparticles have been calculated. This calculation limits the possible range of properties the super particles could possess, such as their mass and lifetime. More positively, it has also been found that if these particles do exist, they will alter the way protons decay in an observable way. Whereas ordinary grand unified theories predict that protons should decay into pions, supersymmetric theories indicate that they should primarily decay into kaons. Since the first observations of proton decay may now be appearing in two separate experiments, it may not be long before the idea of supersymmetry is amenable to direct experimental check. [17]

Supersymmetry is a global gauge symmetry. If it is extended to the more stringent requirement of a local symmetry, then it is called *supergravity* because the symmetries respected by Einstein's general theory of relativity become automatically included. The gravitational force is mediated by a massless boson, the graviton, which because of the supersymmetry, possesses a fermionic partner, the gravitino. Supergravity theory is still in its infancy with no means of direct test yet. But in the future it promises to be a focus of attention because, at present, it is the only obvious route to a superunified theory that can naturally incorporate gravity along with nature's other basic forces. What is also clear from what we already know about supergravity theories is that many of the particles we now believe to be elementary and indivisible must contain internal components if all the symmetries that are demanded by supergravity are to be accommodated. At present, supergravity appears to offer the best hope for unifying all forces and quantizing gravity. It may be, at the very least, a first approximation to a theory that has the power to describe the past, present, and future of the entire physical world.

It seems that what physicists mean by the best, or the ultimate, theories of nature are those that are the most symmetrical. In such theories, there is no diversity in the manifestations of the different forces and fields when they are looked at in the right way. Our ability to unravel the secrets

221

of the universe's conception relies on the fact that as we delve deeper and deeper into the past, we encounter progressively increasing temperature and symmetry. If we look at the universe today, it appears far from its initial symmetric state. The forces of nature are now all distinct; the diversity of galaxies, stars, and planets has replaced the once smoothly distributed, expanding sea of radiation. How is it that the perfect symmetry of the beginning can be so completely disguised today? The world is no longer completely uniform. On the contrary, it favors matter over antimatter, left over right. Why have the perfect symmetries been breached to create the varied and structured world about us?

Why do symmetries get broken?

The breakdown of the symmetry between the different fundamental forces is a manifestation of what is called *spontaneous symmetry breaking*. The intriguing aspect of this phenomenon is that it illustrates how symmetrical laws, embodied in symmetrical equations, can possess asymmetrical outcomes. For instance, the laws governing the motion of a ball rolling down a hill possess no built-in bias for left or right, but perch a ball on the apex of an inverted cone and it will fall in one particular direction. The symmetry of the ruling equations has been broken by a particular outcome. This outcome is dictated by the starting conditions for the ensuing motion, not by the equations predicting the change in the motion with time. The symmetry of the initial state is completely hidden to an observer who looks only at the ball lying finally on one side of the cone or the other. Much of nature is like this. We continually observe asymmetrical outcomes of perfectly symmetrical laws. This is one of the reasons why constructing the past history of the universe is so awkward.

Finally, a human example of spontaneous symmetry breaking. You are at a dinner where the guests are sitting symmetrically around a circular table. Everyone sees a wineglass to their left and to their right (perfect symmetry!). Which one should you pick? The symmetry is broken by the first diner to pick up his wine glass. If a diner reaches to the left, etiquette demands that everyone else follow suit. The dinner party has become left-handed, but neither the laws of human thought nor the laws of human motion are intrinsically left-handed.

Cosmology is a peculiarity of the scientific method. Whereas astrologers strive to predict our futures, astrophysicists are content to probe the past. We cannot manipulate the universe at will and carry out any experiment upon it that we choose. Instead, we must be content to explain *what* is seen by explaining *why* it is seen and to produce as detailed an account of cosmic evolution as we can muster. In this quest, the theoretical ideas with the greatest explanatory and synthesizing power are retained, while arbitrary theories, concocted to explain only one individual phenomenon with no consequences elsewhere, are rejected. Unlike philosophers and writers, scientists have no reason for political or emotional attachment to their theories. They should feel no embarrassment in suggesting mutually exclusive, but possible, explanations and theories. All may be offered as hypotheses to be tested and tried by experiment and rejected only when conclusively disproven.

There is really only one area where cosmologists can make predictions of the future without fear of contradiction at least in their lifetime. It is in the realm of eschatology. We have been continually focusing our attention on reconstructing the birth and early history of the universe, but what of the future. How will space and time end? What fate awaits the dwellers of the far distant aeons of the future?

How will the world end?

"Some say the world will end in fire. Some say in ice." Robert Frost's well-known words still sum up the possibilities fairly accurately. If the universe contains sufficient material then, eventually, the expansion will be reversed into contraction. Redshifts will turn into blueshifts, and the cosmos will plunge into a singularity that will differ in several respects from the one from whence it originated. Whereas the beginning appears to have been regular and quiescent to a high degree, the final state will be chaotic and violent. The present moderate irregularities in the universe will amplify dramatically during the catastrophic approach to the singularity. Furthermore there was a period of "inflation" during the initial expansion, but there will be no "deflationary" phase during the recollapse. The following describes what will happen.

First as the singularity approaches and the temperature rises, all galaxies, stars, and atoms will dissolve into nuclei and radiation. Then the nuclei will be dismembered into protons and neutrons. They, in turn, will be squeezed until the quarks confined within them are liberated into a huge cosmic soup of freely interacting quarks and leptons. At first, there will be more quarks than antiquarks because of the present matter-antimatter asymmetry of the universe, but as the singularity approaches, X bosons will begin to appear. Their growing presence will enable the imbalance between quarks and antiquarks to be repaired. Complete symmetry will be restored. The final plunge into the unknown quantum gravitational era is about 10^{-43} second before the singularity occurs when the density exceeds 10^{96} that of water.

If you do not fancy such a gruesome prospect, even though it lies at least ten billion years in the future, there is an alternative long-range forecast. The following scenario is possible if astronomers fail to find enough cosmic matter to close the universe and make its temporal future finite. Today, just ten billion years after the "big bang," we sit on a hospitable planet in stable orbit around a middle-aged and reliable star. But after another ten billion years elapse, the sun's fuel will be close to exhaustion, and it will expand to encompass the orbit of the earth. Even if we were ingeniously to evade that catastrophe, we would find ourselves evicted from the solar system after about 10^{15} years by the close passage of a neighboring star. Likewise, the sun and its associates will probably be dispatched from our Milky Way galaxy after 10^{19} years. Any stars remaining in galaxies will have completed their steady slide into an all-consuming black hole at the galactic nucleus after about 10^{24} years. Any beings with the resilience and ingenuity to survive all this will still have to cross their greatest hurdle—the decay of all matter. After about 10^{32} years, we expect all protons and neutrons and nuclei to have decayed away. All that will survive are leptons and light, and slowly evaporating black holes. Only after a fantastic 10^{100} years will the black holes that were once galaxies evaporate away, leaving behind unpredictable naked singularities and a sea of inert particles and light. Throughout these long aeons of lingering decay, the shape of the cosmos may change as radically as its contents. The last vestiges of geometrical symmetry will be lost.

If life, in any shape or form, is to survive this ultimate environmental crisis, then the universe must satisfy certain basic requirements. The basic prerequisite for intelligence to survive is a source of energy. Such a source could be present even in the indefinite future, if there were a deviation from complete uniformity in temperature and some degree of disorder. The potential for this does seem to exist. The anisotropies in the cosmic expansion, the evaporating black holes, the remnant naked singularities are all life preservers of a sort. Even when the black holes have all dissolved and the naked singularities are few and far between, irregularities may still grow on a cosmic scale and provide a source of heat as they eventually are smoothed out. An infinite amount of information is potentially available in an open universe, and its assimilation would be the principal goal of any surviving noncorporeal intelligence. As the temperature approaches absolute zero, never quite arriving there, the remaining aeons seem doomed to eternal tedium. But where there is quantum theory there is hope. We can never be completely sure this cosmic heat death will occur because we can never predict the future of a quantum universe with complete certainty; for in an infinite quantum future anything that can happen, will eventually.

Eschatology is nothing if it is not metaphysical. Yet one of the great achievements of modern cosmological theory is that it has transformed the study of the universe from metaphysics into physics. But it would be naive to expunge all metaphysical issues from our discussion just because they might be currently untestable, unmathematical, or simply unanswerable. Cosmology has always been the science closest to theology, and this is undoubtedly the reason for its recurrent fascination to the layperson. It touches upon issues of ultimate significance that are larger than ourselves in a way that the more down-to-earth sciences do not. The biological sciences continually persuade human beings that their position in nature is not in any way special. The harmony between living things and their environments is an inevitable consequence of adaptation. Cosmology, on the other hand, paints a more dramatic picture.

Does the structure of the universe argue for the existence of a Grand Designer?

In many respects, the universe is tailor-made for life. It is cool enough, old enough, and stable enough to evolve and sustain the fragile biochemistry of life. The laws of nature allow atoms to exist, stars to manufacture carbon, and molecules to replicate—but only just. Are all these things coincidences? Should we simply conclude that our universe is not just one of many possible, or even actual, universes, but one of a select subgroup that allows living observers to evolve? This particular universe would necessarily possess the special combination of life-supporting circumstances that are the prerequisites for observers. Or is there but one possible universe and life intimately bound-up with its global structure? Was the cosmos finely tuned to evolve life? The fact that our own universe is unexpectedly hospitable to life is certainly not an inevitable evolutionary effect. The fact that the laws of nature barely, but only barely, allow stable stars to exist with planetary systems today is not a circumstance subject to evolutionary variation. The world either possesses such invariant properties or it does not. A number of independent properties of the universe are so advantageous to the evolution of life that it almost appears designed with our emergence predestined. Could these remarkable "coincidences" be the camouflage of a Grand Designer?

Arguments for the existence of God by appeal to the regularities of nature were last popular with Victorian thinkers, before the discovery of biological evolution and the mechanism of natural selection by Charles Darwin and Alfred Wallace. During this period, people associated the most remarkable feature of the ecosystem, the coincidence that organisms always appeared tailor-made for their environments, with the presence of teleological design in nature. In the inorganic world as well, the invariant rules of physics and chemistry provided persuasive evidence both for design and for a Designer to many great scientists, including Robert Boyle and Clerk Maxwell. What, in retrospect, appears so unusual about the arguments of these times is that we find two incompatible views happily co-existing. One group claims that the evidence for the existence of God lies in the constancy and reliability of the laws of nature, but the other argues that the prime evidence for a deity lies in reports of miracles—breakdowns in the laws of nature.

Anyone looking at the modern findings about the universe faces a similar paradox. On the one hand, there is a striking symmetry underlying the universe, and were it not present, life could not exist. At the same time, we see that these symmetries are invariably only "almost symmetries," and the tiny violations of perfect symmetry that we observe are equally necessary for our existence. It is the minute deviation from complete uniformity in the universe that allows the existence of galaxies, planets, and people. It is the slight imbalance between matter and antimatter that allows matter

Conclusions and Conundrums

to survive the extremes of the "big bang"; without it the universe would contain only radiation. It is these tiny, fortunate breaches of complete symmetry that are perhaps what cosmologists would today call "miracles." Someday we may understand them in terms of more basic necessities, but, either way, they allow us to do nothing more than reinforce the views we already possess about a Grand Designer. There is certainly no unique interpretation of the universe we witness.

The question of the precise identity of any such Grand Designer has always been a problem for any advocate of a cosmological design argument.

Religious thinkers have often appealed to cosmology for evidence of a Creator. We have seen, however, that in cosmological theories, the role of Creator is essentially assumed by the naked "big bang" singularity. Anything, we believe, can emerge from such a singularity unless there exist some rules unknown to us that control its behavior. Does a naked singularity have any other characteristics of a Creator? Fortunately not, for man could, in principle, one day make his own local naked singularity by moving so much matter into a confined region of space that a black hole appears. A naked singularity may lie within it. The theological status of naked singularities must be dramatically lowered if they can be man-made. The act of creation, as evidenced by matter spewing forth from a naked singularity, may still require a Creator. Naked singularities do not qualify as deities. If man could instigate a naked singularity, he would also, incidentally, possess the means to destroy space and time! Fortunately, there is little likelihood of politicians adding naked singularities to their arsenals of destruction in the present and forthcoming centuries.

Rather than pursue these unbridled speculations any further, let us return to consider some more practical issues, ones that we can hope to resolve with the aid of twentieth century technology.

Will the main cosmological questions be answered by the year 2000?

New space probes will expand the astronomical frontiers along the entire electromagnetic spectrum. With the ensuing flood of new information, we should certainly be able to decide whether Hubble's constant appearing in the universal red shift versus distance relation is close to 50 rather than 100 kilometers per second for each megaparsec of distance. But will we determine whether the universe is open or closed? Here one has cause for pessimism. In the past, practically every advance in instrumentation has actually hindered progress towards this particular cosmological goal. Instead, we have learned a lot about galactic evolution and the subtleties of the large-scale structure of the universe. Also, it is quite possible, and even predicted by inflationary theories, that the universe could be expanding at almost exactly the critical rate that divides the open and closed universes. If that is so, then the universe might expand forever, but we will never be able to decide by observation whether or not this is the case.

As for the origin of galaxies, here one has room for genuine optimism, although a theory of quantum gravity will be necessary before the issue can be resolved. Progress in theoretical high-energy physics has been so rapid during the last decade that a quantized gravitation theory, unified with theories of other fundamental forces, seems at last to lie within reach. Whether any way can be conceived to test it by experiment is another matter. There is no principle of cosmic convenience which guarantees that such theories can be tested with twentieth-century technology. Indeed, they may not be testable at all. Nevertheless it is only when physical theories have undergone this radical extension that it will be possible to talk meaningfully about events occurring during the first 10^{-43} second of the universe's expansion history. It is at that moment that the universe's ultimate secrets are undoubtedly hidden.

One last thought should offer us encouragement. In the past, the most exciting and far-reaching discoveries in astronomy have always been unexpected—quasars, pulsars, the microwave background radiation, the binary pulsar, bursting x-ray sources. Moreover, the stream of new theoretical ideas and the serendipity of new experimental discovery show no signs of abating. Knowledge appears like an expanding ball, ever growing in volume but at the same time ever increasing the extent of its boundary with the unknown that awaits beyond its edge.

Could there be any short cuts to the answers to the cosmological questions? There are some who foolishly desire contact with advanced extraterrestrials in order that we might painlessly discover the secrets of the universe secondhand and prematurely extend our understanding. Such a civilization would surely resemble a child who receives as a gift a collection of completed crossword puzzles. The human search for the structure of the universe is more important than finding it because it motivates the creative power of the human imagination. About 50 years ago a group of eminent cosmologists were asked what single question they would ask of an infallible oracle who could answer them with only yes or no. When his opportunity came, Georges Lemaître made the wisest choice. He said, "I would ask the Oracle not to answer in order that a subsequent generation would not be deprived of the pleasure of searching for and finding the solution."

Notes

1. A novel technique for inferring "peculiar" velocities, that is to say, velocities of galaxies relative to the cosmic expansion, has led to the realization that the Virgo cluster of galaxies, known from its redshift to be at a recession velocity of about 1000 km/s, has a substantial component of peculiar velocity. Virgo is moving towards a complex of galaxies about four times further away in the direction of Centaurus, dubbed the "Great Attractor", with a peculiar velocity of about 400 km/s. Recession velocity is a convenient way of measuring distance: every increase of 1000 km/s in recession speed is equivalent to an increase in distance away of between 10 and 20 Megaparsecs, depending on the choice of the Hubble constant.

2. Cygnus Xl is the oldest and most studied black hole candidate, there are a number of others. For a good account of their properties and the astronomical aspects of black holes in general see the recent book by Luminet listed under *Further Reading*.

3. Johnstone-Stoney derived his units of mass length and time by combining G, c and e, the electric charge of an electron. This produces natural units that differ from Planck's only by a small numerical factor equal to the square root of 137.

4. The most publicised developments in this direction have been those of James Hartle, Stephen Hawking and others which attempt to develop a description of the entire Universe as a quantum entity. This leads to new possibilities for the beginning of the Universe. In some cases it is possible to interpret the mathematical description of a quantum universe as coinciding with that expected if the Universe had come into being spontaneously out of nothing. Another possibility that has been explored, which is described in 'A Brief History of Time' by Stephen Hawking is that the nature of time itself was radically altered in the high-density environment of the Big Bang. In this picture a reconstruction of the Universe's past eventually enters a regime where one finds time has become another dimension of space rather than the origin of time, which a reversal of the universal expansion would lead one to expect. (See *Theories of*

Notes

Everything by J. D. Barrow, listed in *Further Reading*, for a fuller discussion of these ideas and some of their wider implications.)

5. In the last three years there has been enormous interest in the issue of what happens to all the information stored in configurations of matter if a small black hole were to form from it and then evaporate into radiation. A vast amount of information is needed to specify an object like a star completely. This information cannot be carried off by the particles that are radiated from the black hole during the evaporation process. If the black hole evaporates completely into radiation what happens to all the original information? Is it lost from the Universe in some mysterious way? If so, something new is happening that existing quantum physics does not describe because information can neither be created nor destroyed when quantum systems change. At present, this remains an important unsolved problem about black holes. It may be that new physics is needed that will show the information that goes into the black hole is indeed destroyed. Or, perhaps this new physics will reveal that the information is really hidden in the radiation evaporated from the black hole. Alternatively, the black hole may never completely evaporate. It may leave behind a stable relic object with a mass very close to the Planck mass. This relic would have to be a very strange object indeed, far more complex than anything else we have encountered or conceived of in Nature, because it would have to act as a repository for all the information required to specify the structure of a star at least 10^{43} times bigger.

6. These earlier reported detections of free quarks could not be repeated in later experiments and were not confirmed by other experimental searches. All the available evidence shows that quarks are confined.

7. It has been shown that the matter-antimatter asymmetry of the Universe might also be established by similar asymmetrical decays of quarks and antiquarks when the Universe is much cooler, about 10^{15} degrees Kelvin.

8. Proton decay has still not been seen. The best grand unified theories predict that the proton lifetime should be longer than 10^{32} years and this is beyond the reach of existing experiments to detect. By a remarkable piece of good fortune it was the large underground detectors, built for the purpose of seeing proton decay, that detected neutrinos from the nearby supernova SN1987A that was seen to explode near our galaxy in 1987.

9. Lithium is the latest addition to the menagerie of cosmic relics. This light element is produced in significant amounts in the first few minutes of the Big Bang, and is also generated by collisions between cosmic rays and interstellar atoms of carbon, nitrogen and oxygen. By searching for lithium in the oldest stars, a minimal abundance has been found that is suggestive of a Big Bang origin. The combination of helium, deuterium and lithium abundances leaves little freedom in the cosmological model if it is to account for the data in terms of a pregalactic, primordial origin. Also, when these element abundances are explained, the density of baryons is tightly constrained. This gives cosmologists an important indicator of the total density of baryons in the Universe regardless of whether they are luminous or dark today.

Novel experiments are searching for baryonic dark matter in the halo of our galaxy. The dark baryons are likely to be the remnants of stars, or objects of planetary mass, too low in mass to have formed stars. If the dark matter consists of massive objects anywhere in the range between 10^{-6} and 10 solar masses, there will be a gravitational lensing effect that temporarily brightens background stars. As a dark object, orbiting in our halo, intersects the line of sight to a distant star, its gravity field acts like a lens and amplifies the starlight. The effect is weak, and one expects to examine millions of background stars to see a single gravitational amplification event. Several experiments underway in the period 1993/95 at the European Southern Observatory in La Serena

Notes

and the Carnegie Observatory in Las Campanas, Chile, and at the Mount Stromlo Observatory in Australia should find conclusive evidence for Massive Astrophysical Compact Halo Objects, or MACHOs, if they exist and constitute halo dark matter in our galaxy. Preliminary results from these experiments were reported in late 1993: the first three MACHO events were found in surveys of more than three million stars in the Large Magellanic Cloud. The duration of the event depends only on the mass of the MACHO, its distance away, and its speed. Since the MACHOs are assumed to be in the halo of our Galaxy, where their orbital velocities are known, one can infer that they typically have a mass about one-tenth of the mass of the Sun, but with at least a factor of three uncertainty because of the paucity of data. This is never the less an intriguing result. More data are reguird to tell whether MACHOs are responsible for all of the dark matter in our Galaxy's halo and to discover whether their masses correspond to those of giant planets or defunct stars.

10. The study of the process of inflation has grown into a major branch of theoretical cosmology. A phase transition in the state of matter is just one way in which it can occur. The most important aspect of these models has been their ability to explain the origin of the small variations in the density of matter from place to place in the Universe which give rise to galaxies. Inflation takes the natural microscopic fluctuations that must exist because of quantum randomness and stretches their extent up to the astronomical dimensions of galaxies and clusters of galaxies. The COBE satellite may have detected these fluctuations in their early stages, before they condensed into galaxies, and the observed pattern of fluctuations is so far found to be in agreement with the particular pattern of variation predicted by most inflationary universe models.

11. Temperature fluctuations of about 1 part in 10^5 in the cosmic microwave background over degree scales are predicted, if structure has indeed grown by gravitational instability from primordial density fluctuations laid down at the epoch of inflation. The COBE satellite measured temperature fluctuations over 10 degree scales, that correspond in magnitude to the prediction of the inflationary cosmological model. These scales are larger than those which gave rise to the structures we observe in the large-scale galaxy distribution. The fossil counterpart fluctuations that seeded the great superclusters of galaxies and voids are degrees across on the sky. Many experiments are being designed to search for these smaller-scale fluctuations. At the present time, we can hope that confirmation of such fluctuations is imminent, although not yet confirmed because of the complex foreground emission from our galaxy. Experimentally, it is rather like looking through a dirty windshield. COBE succeeded, in large part because it surveyed the entire sky and flew above our atmosphere where some of the confusing noise originates, but the challenge confronting the small-scale ground-based experiments is more difficult.

12. Supernova 1987A in the Large Magellanic Cloud provided brilliant testimony to the supernova theory of stellar core collapse to a dense ball of compressed neutrons. The neutrons form by the reaction:

$$\text{proton} + \text{electron} \Rightarrow \text{neutron} + \text{neutrino.}$$

The weakly interacting neutrinos pass freely out through the outer layers of the collapsing stellar core at the speed of light, and were detected in independent experiments performed deep underground in the U.S. and in Japan. The experiments were run in deep underground mines to avoid spurious cosmic ray signals and were searching, unsuccessfully, for evidence of proton decay. The neutrino flux at the earth was consistent with that expected in the process of converting the iron core of an evolved star in its final death throes to a neutron star. This was the first detection of a source of neutrinos from outside our solar system.

235

Notes

13. The rapid, accelerated expansion of the Universe during a period of inflation ensures that the present-day Universe will be highly isotropic so long as the period of inflation lasts long enough. While the expansion accelerates, all the distorting influences on the expansion rapidly become insignificantly small, leaving the expansion very close to an isotropic state. The only deviations from perfect isotropy and uniformity are created by the tiny (one part in 10^5) quantum fluctuations in the density of matter that inflation produces. Thus we see that inflation achieves the aims of the chaotic cosmology programme by explaining the present uniformity and isotropy of the Universe irrespective of its initial state. However, whereas the original chaotic cosmologists imagined that this would be done through dissipating irregularities and anisotropies by frictional processes, inflation does not dissipate any irregularity. It simply sweeps it far beyond the horizon of our visible part of the Universe. On this picture, the visible universe owes its observed isotropy and comparative uniformity to the fact that it is the expanded image of a part of the early Big Bang that was small enough to be kept smooth by physical processes. By contrast, in the absence of a period of inflation the Universe expands too slowly early on for the visible part of the Universe today to have arisen from a region small enough to be kept smooth by physical processes acting at rates less than the speed of light. The Universe expands faster than the speed of light during its early expansion history. This does not create any conflict with Einstein's theories of relativity. They only require that the signalling of information not exceed the speed of light. The expansion involves no signalling or transmission of information at a speed faster than light despite the fact that separate parts of the Universe do separate from each other faster than the speed of light.

14. The smallness of the cosmological constant remains a great mystery. Physics seems to demand that its value be about 10^{120} times bigger than astronomical observations allow it to be. Clearly we are lacking knowledge of some important feature of the laws of Nature which ensures that its value is suppressed. One interesting possibility introduced by Sidney Coleman (see Barrow, *Theories of Everything* in the list of Further Reading for a description) is that space-time might have an intricate crennellated structure with handles and loops joining back upon itself, rather than being smooth and relatively featureless. These structures are all about the Planck length in size. If this is the case then the array of loops and handles gives rise to fluctuations that will almost exactly cancel out any cosmological constant that pre-exists. This leads one to predict that, with high probability, the cosmological constant will have a very small value.

15. This one-off event has neither been repeated nor definitely explained. No magnetic monopoles have been detected and this is what one would expect if inflation had occurred in the early universe.

16. Particle physicists discovered that the problems of consistently joining the electroweak and strong forces of Nature with gravity could be overcome by relinquishing the assumption that the most basic entities are point-like. 'String' theories, in which the most basic entities are lines of energy, turn out to require the unification of the other forces of Nature with gravity for their logical consistency. If we imagine the basic structures in these theories to be tiny elastic loops then strings behave in such a way that the tension in the elastic slackens as the temperature increases but tightens as it falls. Consequently, as the temperature falls to that of terrestrial experiments the loops contract and become increasingly point-like. This ensures that they reproduce all the successful predictions of the point particle theories. However, when the temperature rises to the Planck level the string tension is small and the loops become intrinsically string-like in their behaviour. One of the appeals of superstring theories is their potential to explain the masses of all the elementary particles. The point particle

Notes

theories were faced with coming up with explanations for the masses of each of these. But a single string might be able to explain them all. Any string, like a guitar string, has a set of natural frequencies at which it will vibrate. Each of these possesses a different energy. In the case of the natural vibrations of a superstring, these energies each correspond (via Einstein's $E = mc^2$ mass-energy equivalence) to the masses of different particle-like states when the tension is high. So determination of this rainbow of superstring vibrations might one day tell us the masses of the stable elementary particles of matter. Most of the vibrations of the string will be associated with energies close to the Planck energy and with short-lived heavy particles that disappeared in the first moments of the Big Bang. So far, string theories have proven easier to find than to solve and physicists have been unable to solve the mathematical equations and extract the predictions of string theory for the early evolution of the Universe. There seem to exist many different, logically consistent string theories. Finding which, if any, of them describes our Universe remains a formidable challenge for the future.

String theories are also known as 'superstring' theories to emphasize the fact that the strings possess supersymmetry. Superstrings should be distinguished from cosmic strings (p. 203) which may or may not be a consequence of superstring theories.

17. Existing particle accelerators have still found no evidence of supersymmetry. They are not powerful enough to reach the energy scale of several thousand times the proton's mass energy where supersymmetry will be explicitly manifested. The great emphasis placed by the particle physics community upon the need to build the next generation of 'supercolliders' is motivated in part by the desire to discover direct evidence of supersymmetry. It is also possible that the WIMP detectors being built to detect weakly interacting dark matter particles in our galaxy (see p. 128) might be the first experiments to find evidence for supersymmetry.

Glossary

adiabatic fluctuations. The spatial variations in the density of the universe in which both the matter and radiation densities vary from place to place simultaneously.

anisotropy/anisotropic. A dependence on direction; a lack of isotropy. An anisotropically expanding universe would expand at different rates in different directions.

anthropic principle. An idea that the universe possesses many of its extraordinary properties because they are necessary for the existence of life and observers.

antimatter/antiparticle. A particle with identical mass and spin as another particle, but with equal and opposite values of other properties like electric charge, baryon and lepton number, and so on. Every particle possesses an antiparticle although some electrically neutral particles, like the photon and the pion, are their own antiparticles. When a particle encounters its own antiparticle both annihilate into radiation. The antiparticle of the neutrino is called the antineutrino, that of the proton the antiproton, and so on.

asymptotic freedom. A property of interactions between quarks, that the forces between them become progressively weaker at high temperatures and small distances of separation.

baryons. A class of elementary particles, including the neutron and the proton, which take part in strong interactions.

baryon number. The baryon number of a system is the sum of the total number of baryons minus the total number of antibaryons.

baryon to photon ratio. The average number of baryons (protons) to photons in the universe. Today this is close to one in a billion.

Glossary

"big bang." A standard model of the universe in which all matter, space and time expand from an initial state of enormous density and pressure.

black body radiation. Heat radiation which is in equilibrium with matter and so absorbs and emits the same amount of energy at any wavelength. The variation of intensity with frequency has a characteristic shape called the Planck spectrum.

black hole. A gravitationally collapsed body from which nothing, including light, can escape. An external observer can measure only three properties of a black hole: its mass, electric charge and angular momentum.

blue shift. The shift of spectral lines toward shorter wavelengths in the spectrum of an approaching source of radiation.

boson. A class of elementary particles with integer units of the basic unit of spin $h/2\pi$; examples are the photon, X, W, and Z bosons. More than one boson can exist with identical quantum numbers. They do not obey the Pauli Exclusion Principle.

chaotic cosmology. The idea of supposing the universe began in a highly chaotic state but smoothed itself out into the present ordered state in the course of time by frictional processes.

closed universe. A model of the universe that is finite in total volume and in total age. It evolves from a "big bang" to a point of maximum expansion before contracting back to a "big crunch" of high density and temperature.

color. In elementary particle physics color is a property labelling different types of *quark.* It has nothing to do with visual color. The strong interaction influences particles possessing the color attribute, that is, quarks and gluons.

Coma cluster. An aggregate of thousands of galaxies some 300 million light years from us and millions of light years in extent. There are many other similar clusters of galaxies in the universe.

confinement. A property of quarks; the force between two quarks increases with their separation and may prevent them being observed individually. They are confined in triplets inside baryons or in pairs within mesons at low energy.

conservation law. A rule which states that the total value of some particular quantity is unchangeable in physical interactions. Quantities, like energy and electric charge, which obey conservations laws are called conserved quantities.

coordinate singularity. A place where the coordinate lines of a mapping system degenerate and intersect; for example, in the case of latitude and longitude lines, at the North and South poles of a terrestrial globe.

cosmic censorship. A hypothesis suggested by the English physicist Roger Penrose that singularities in space and time are always surrounded by event horizons which prevent them being observed, and stop them from influencing the outside world.

cosmological constant. A term added by Einstein to his original gravitation theory in 1917 to enable it to predict a static, non-expanding universe. There is no evidence that such a quantity exists as yet, although it can also be present in a non-static universe and may be responsible for unusual behavior in the first instant of the universe called inflation.

Glossary

critical density. If the average density of the universe exceeds this value, about 2.10^{-29} grams per cubic centimeter, it will recollapse in the future. If it does not exceed this value it will expand forever.

curved space-time. According to Einstein's general theory of relativity, gravity causes a distortion of the space and time fabric of the universe; the paths of light rays and the rates of clocks are influenced by the presence of masses as though they were in a curved, non-Euclidean geometry.

de Sitter universe. A model of an expanding universe which contains no matter, but the cosmological constant acts as a long-range repulsive force causing distant parts of the universe to recede and expand at a rate far faster than in Friedman's model universes. The universe is not like this today, but may once have been during an early period of inflation 10^{-35} second after the "big bang."

deuterium. A heavy isotope of hydrogen with a nucleus composed of one neutron and one proton. It has similar chemical properties to hydrogen. The abundance of deuterium in interstellar space is about 2.10^{-5} that of hydrogen.

disk galaxy. A flat pancake-shaped galaxy with a radius up to about twenty-five times that of its thickness.

Doppler shift/effect. The displacement of spectral lines in the radiation received from a source due to its relative motion along the line of sight. A motion of approach results in a blue shift; a motion of recession results in a red shift.

electron. The lightest massive elementary particle that is (negatively) electrically charged. It shows no evidence of possessing any internal structure or constituents and has a mass of 9.10^{-28} gram.

electron volt. The energy acquired by an electron when accelerated by a potential difference of one volt; abbreviated eV; one electron volt $= 1.602 \times 10^{-12}$ erg.

elliptical galaxy. A galaxy whose structure is smooth and amorphous, without spiral arms, and ellipsoidal in shape. They range in mass from 10^7 to 10^{13} solar masses and are redder in color than spiral galaxies of comparable mass.

entropy. A measure of the amount of disorder in a system; it never decreases in any physical interaction.

entropy per baryon. A measure of the relative distribution of energy in disordered and ordered forms in the universe. It is given by the number of photons per baryon in the present universe and is roughly equal to one billion, the reciprocal of the baryon to photon ratio.

escape velocity. The minimum velocity that must be attained to completely escape the gravitational pull of a body. It depends on the mass and radius of the body.

event horizon. The boundary of a black hole. No signals or particles can travel from inside the event horizon to the outside.

fermion. An elementary particle with half integral units of the basic unit of spin, $h/2\pi$; examples are electrons, protons and neutrinos. No two fermions can exist with identical values of the quantum numbers characterizing them.

flat universe. A model universe possessing the largest density (the critical density) that allows it to continue expanding forever. It is so called because at any instant the global geometry of space is Euclidean, space is flat rather than curved.

gamma rays. Photons of very high energy and short wavelength; the most penetrating form of electromagnetic radiation.

241

Glossary

gauge invariance theories. A very successful class of theories for the weak, electromagnetic, strong and gravitational forces. Such theories have a form that is dictated by the requirement that they are unchanged by a symmetry transformation whose effect varies from place to place.

GeV. Gigaelectronvolt = 10^9 electron volts.

gluon. A particle of zero mass possessing color; quarks interact through the exchange of gluons. There are eight varieties of gluons.

grand unification/unified theories (GUTs). A class of gauge theories that unite the strong, electromagnetic and weak interactions at high energy. Ultimately, it is hoped that they can be extended to incorporate gravity.

gravitational instability. The process by which a lumpy medium exceeding a certain critical size (the Jeans length) grows lumpier with time because of the gravitational attractions of its constituents; it is thought to be the reason why galaxies exist.

gravitino. A hypothetical fermion predicted by supersymmetric gauge theories. It has spin 3/2 units and a non-zero but, as yet, unpredictable mass.

graviton. It is massless and has a spin of two units of the basic spin, $h/2\pi$. The graviton is a boson.

hadron. Any particle that participates in strong interactions. Hadrons contain *quarks* and are divided into two classes: baryons (which are fermions) and mesons (which are bosons).

halo. The diffuse, nearly spherical cloud of old stars and globular star clusters that surrounds a spiral galaxy.

heavy elements. In astronomical parlance this term is used of nuclei heavier than boron. It includes the important biological elements like carbon, nitrogen and oxygen.

helium. The second lightest and most abundant element in the universe. There are two stable isotopes of helium (He): He^3 and He^4. He^4 contains two protons and two neutrons in its nucleus while He^3 has one less neutron. The He^4 nucleus is the alpha particle of radioactive decay. One quarter of the universe's mass is in the form of He^4.

horizon. The observable region of the universe; at any time this is dictated by the distance light can have travelled since the origin of the universe.

Hubble's constant. The present expansion rate of the universe; it is the constant of proportionality linking the recession velocity to the distance of galaxies. Observers disagree on its value within the range 50 to 100 kilometers per second per megaparsec.

inflation/inflationary universe. The period during the first 10^{-35} second of the universe when a phase transition accelerates the expansion rate according to some grand unified gauge theories.

isothermal fluctuations. The variations in the density of the universe from place to place in which the radiation density remains constant while the matter density changes.

isotropy. Independence on direction or angle. In an isotropic universe, all measurable quantities are the same in any direction.

Glossary

Jeans mass/length. The critical size above which a smooth medium inevitably degenerates into a clumpy one because of gravitational attractions. This tendency can be resisted by thermal pressure in systems smaller than the Jeans length.

leptons. A class of elementary particles that do not participate in strong interactions; it includes the electron, muon, tauon and the neutrinos. Leptons show no evidence of any internal structure. The lepton number is the total number of leptons minus the total number of antileptons in a system.

light year. The distance travelled by light rays in one year, equal to 9.46×10^{17} centimeters.

Local Group. The small system of galaxies to which our Milky Way galaxy belongs. The largest member is Andromeda; the Milky Way is the second largest member and there are twenty or more other, smaller member galaxies.

magnetic monopole. A massive particle predicted to exist by grand unified theories. It possesses internal structure and a mass near 10^{-8} gram along with the long range magnetic field of a single isolated magnetic pole.

megaparsec. one million parsecs; a parsec is 3.086×10^{18} cm which is 3.26 light years; abbreviated Mpc.

megaelectron volt. One million *electron volts;* abbreviated MeV.

meson. A class of strongly interacting particles including pions and kaons which have zero baryon number. They are bosons.

microwave background. The universal radiation field at microwave frequencies with a black-body spectrum corresponding to heat radiation at about three degrees Kelvin.

mini black hole. A small black hole with a mass much less than that of the sun (2.10^{33} gram) formed by the pressures of the very early universe rather than by the death and collapse of an ordinary star.

Mixmaster universe. An extremely anisotropic model of the first moments of the universe's expansion. This universe would expand at different rates in two perpendicular directions while undergoing collapse in the third. The volume would increase but the directions of expansion and collapse are subsequently permuted, and change many times in the course of expansion.

model universe. A mathematical idealization of the universe derived from a physical theory of how gravity acts. The most successful such gravity theory is general relativity.

muon. An elementary particle with negative electronic charge, similar to the *electron* but 207 times heavier. It is a lepton.

naked singularity. A point where the laws of physics break down that is not hidden to distant observers by an event horizon.

neutrino. An electrically neutral particle having weak and gravitational interactions only. The electron and muon type neutrinos, along with their antiparticles, are known to exist but a third tau neutrino should exist and possibly many others. Neutrinos are leptons and also fermions.

neutrino viscosity. A process that can arise when the universe is about one second old. Anisotropies in the cosmic expansion and very small non-uniformities can be smoothed out by the motion of neutrinos over large distances.

Glossary

neutron. An elementary particle with zero electric charge about 1838 times heavier than the electron. It is a fermion and a baryon and participates in strong, weak and gravitational interactions.

neutron star. A star whose core is composed primarily of neutrons, with a mean density of 10^{14} gram per cubic centimeter. Pulsars are rotating neutron stars.

Newton's constant. The fundamental constant first introduced by Isaac Newton to characterize the intrinsic strength of gravity; denoted by G. The gravitational force between two bodies is G times the product of their masses divided by the square of the distance between their centers, $G = 6.67 \times 10^{-8}$ cm^{3}gm^{-1}s^{-2}.

"no hair" theorem. A mathematical argument proving that black holes can possess at most three observable properties accessible to outside observers; their mass, electric charge and angular momentum. Two black holes with equal values of these three attributes are completely indistinguishable; they possess no other defining characteristics.

normal distribution. Frequency of different outcomes for a completely random series of events first found by Gauss. The form of the distribution does not depend on the details of the actual physical process generating the random events.

nucleon. A component of an atomic nucleus, either a proton or a neutron.

Occam's razor. A guiding principle in developing scientific theories that requires one to prefer an explanation that uses the minimum number of hypotheses.

Olbers' paradox. The question "Why is the universe dark at night if the universe is infinite?"

open universe. A model of the universe which will expand for all future time; so called because the geometry is that of a negatively curved space.

pancake. A term used by astronomers to describe the form of embryonic galaxy clusters that collapse rapidly in one direction to form huge, flattened sheets of galaxies.

parsec. 3.26 light years.

Pauli exclusion principle. The principle that no particles can possess precisely the same set of quantum labels. This principle is obeyed by fermions but not by bosons.

phase transition. A sharp transition of a system from one equilibrium state to another, usually accompanied by a change in symmetry. Examples are freezing, melting and boiling in the case of water. Elementary particle states can also undergo analogous changes accompanied by symmetry and energy changes.

photon. A particle associated with light. It has zero mass and is a boson. It mediates electromagnetic forces between electrically charged particles.

photino. A hypothetical fermion arising in some supersymmetric theories. Its mass is non-zero but not precisely known.

pion. Also called the pi-meson. It is a hadron that can be either positively or negatively charged or electrically neutral.

Planck's constant. The fundamental constant of quantum mechanics denoted by h, equal to about 6.625×10^{-27} erg second.

Planck time. 10^{-43} second, the moment before which Einstein's theory of general relativity must be improved to include quantum theory. It can be expressed as

Glossary

$(Gh/c^5)^{1/2}$ where G is Newton's gravitation constant, h is Planck's constant and c the velocity of light.

positron. The antiparticle of the electron. It possesses a positive electric charge.

proper time. Time measured by a clock sharing an observer's motion. Clocks in relative motion with respect to an observer will measure his time to flow at a different rate to proper time.

protogalaxy. A precursor state to the present one for a galaxy when it was predominantly gaseous and undergoing dynamical and chemical evolution. The protogalaxy phase lasted for up to a billion years or so.

proton. A positively charged particle found in atomic nuclei and having a mass 1836 times larger than that of the electron. It is a fermion and also a baryon and feels the strong, electromagnetic, weak and gravitational forces. Protons possess three quarks as internal structure and so are not elementary particles.

quantum theory. The fundamental theory of micro phenomena and particles wherein they are found to possess both wave and particle characteristics.

quark. The fundamental particles of which hadrons are composed. Quarks have a charge equal to one-third or two-thirds the electron charge and possess color.

quasars. A class of astronomical objects that appear starlike but whose energy emission is several billion times larger than that of a star. They have large red shifts, up to about 3.5, and are the most luminous objects in the universe.

radiation era. That period of the universe's history, lasting until it is about 100,000 years old, during which the density of radiation exceeds that of matter and the universe was in an ionized plasma state. No atoms, stars, planets or galaxies existed.

recombination. The capture of an electron by a positive ion to form a neutral atom. This process occurs universally in primordial matter when the universe cools to about 3000 degrees Kelvin, after about one-third of a million years of expansion. Thereafter the cosmic radiation is essentially collisionless.

red shift. The Doppler shift of spectral lines towards longer wavelengths in the spectrum observed from a receding source of radiation.

relativistic particles. Particles whose velocities are equal or close to the velocity of light, 3.10^{10} centimeters per second.

rishons. Hypothetical internal constituent of quarks and leptons that come in two types, T rishon with electric charge 1/3 and the V rishon with zero charge. The antirishons have equal and opposite charges. All elementary quarks and leptons can be constructed as combinations of these particles.

singularity. A portion of the edge of the universe, possibly where the density and temperature are infinite as at the "big bang." The laws of physics break down at a singularity.

singularity theorems. A collection of precise mathematical arguments that prove that a universe will contain a singularity in the past if a number of specific assumptions about its structure are true.

sleptons. Hypothetical elementary particles present in supersymmetric gauge theories. They are bosons and there will be one type for each lepton.

spheroidal galaxy. A galaxy with the shape of a slightly squashed sphere. The earth is almost spheroidal in shape.

Glossary

spin. Fundamental intrinsic property of elementary particles which describes the state of rotation of the particle. The spin is always found to be in whole or half units of Planck's constant divided by 2π. Particles with whole spins are called bosons, those with half spins are called fermions.

spiral galaxy. A galaxy with a prominent central bulge embedded in a flat disk of gas, dust and young stars that wind out in spiral arms from the nucleus. The Milky Way is a spiral galaxy.

spontaneous symmetry breaking. A sudden change in the equilibrium state of a system; for example, when a pencil balanced on its point falls. The equilibrium of elementary particle states can change in a similar way to hide the symmetry of the starting configuration.

squarks. Hypothetical elementary particles that exist in some supersymmetric gauge theories. They are bosons and partner the quarks.

steady-state cosmology. The cosmological theory of Hermann Bondi, Thomas Gold, and Fred Hoyle in which matter is continuously created to fill the voids left as the universe expands. Consequently, the universe should have no beginning and no end and should always maintain the same mean density. The required creation rate amounts to adding one atom to every cubic meter of space every ten billion years. This theory is now in conflict with observations and has been abandoned by cosmologists.

strings. The tubelike configurations of energy that can arise in the early moments of the universe. Typically, a string would have a thickness of only 10^{-27} centimeter, but if extended across the entire observable universe would have a mass of about 10^{17} suns.

strong interaction/force. The strongest of the fundamental forces of nature. It is responsible for binding the atomic nucleus together and has a very short range of about 10^{-13} centimeter. It affects all hadrons but neither leptons nor photons. The strong force also acts between particles like quarks and gluons which possess color and it is then often referred to as the color force.

supercluster. A cluster of galaxy clusters, about 10^8 light years in extent.

supernova. A cataclysmic explosion of a star in which the outer portions of the star are blown off and the inner core compressed. A supernova produces more energy in a few days than the sun has radiated in a billion years.

supersymmetry. A property demanded of certain unified gauge theories which creates a symmetry between fermions and bosons. If this symmetry is a local gauge symmetry it can include the properties of the gravitational interaction and the resultant theory is called supergravity. These theories are now being actively investigated by theorists.

tau lepton or *tauon.* An elementary particle with negative electric charge. It has a mass of 3491 electron masses, and is a charged lepton like the electron and muon.

thermal equilibrium. The steady-state attained by a system that is in close contact with a thermal reservoir at some constant temperature.

topology. Concerned with the intrinsic properties of shapes, surfaces, and spaces together with the possible configurations in which they can exist.

trapped surface region. A region of space and time from which light rays cannot escape because of the restraining influence of gravity upon them.

Glossary

uncertainty principle. Enunciated by Werner Heisenberg, it shows that one cannot precisely determine both the position and the velocity of a particle at the same time. Also if a system exists for a finite time there is a limit, given by Planck's constant, to the accuracy with which its energy can be measured even with perfect instruments.

vacuum. The lowest energy state of a physical system.

vacuum fluctuation. As a consequence of the uncertainty principle, pairs of particles and antiparticles are spontaneously appearing and disappearing in space and time after existing for an unmeasurably short time. These created pairs are called vacuum fluctuations.

vacuum polarization. If *vacuum fluctuations* occur in the neighborhood of an electrically charged particle, the members of the created pairs with opposite charge to that particle will be preferentially attracted towards it. This migration is called vacuum polarization.

velocity of light. 3.10^{10} centimeters per second in empty space.

virtual particle. A component of a vacuum fluctuation that cannot be observed directly.

viscosity. The internal friction of a fluid or medium that tends to resist and dissipate its motion and any irregularities within it.

void. A large region of the universe devoid of luminous galaxies.

wall. A sheetlike configuration of energy that can arise in the very early universe according to some gauge theories. The thickness of the wall will be about 10^{-27} centimeter. Observations rule out the existence of a wall stretching across the entire observable universe.

wavelength. In any kind of wave, the distance between the peaks of successive wave crests.

W boson. Recently discovered electrically charged elementary particle with a mass close to 80 GeV that is predicted to exist by the gauge theory unifying the weak and electromagnetic forces. It was discovered in 1983.

weak interaction/force. One of the fundamental interactions experienced by elementary particles. It has a very short range of 10^{-15} centimeter and is responsible for radioactivity.

whimper. Space-time *singularity* not accompanied by infinities in physical quantities like density or temperature.

white dwarf. A compact star of mass about that of the sun but size about that of the earth. Within a white dwarf, electrons are stripped from the outer regions of atoms by the force of gravity and move freely. Their motion creates a pressure that resists the crushing force of gravity. White dwarfs have been observed.

white hole. A hypothetical time reverse of a black hole out of which material would continuously appear at the velocity of light. A local example of the "big bang," but there is no evidence for their existence.

white noise. A completely random and uncorrelated noise, with equal power at all frequencies.

wino. A hypothetical supersymmetric relative of the W boson. It is a fermion.

X boson. A very heavy analogue of the photon predicted by grand unified gauge theories and which mediates transmutations between quarks and leptons. It

Glossary

mediates the decay of protons for which there is some recent evidence awaiting confirmation.

x-ray. Light with wavelength between 10^{-9} and 10^{-7} centimeter. The second most penetrating form of radiation after gamma radiation.

Z boson. Electrically neutral companion particle of the W boson predicted to exist in gauge theories of the weak and electromagnetic forces. It has a mass close to 93 GeV and was discovered in 1983.

zino. A hypothetical supersymmetric partner to the Z boson. It is a fermion.

Further Reading

Popular

Barrow, J.D. Theories of Everything, Oxford UP, Oxford, 1991; Ballantine, N.Y., 1992.
Carrigan, R.A. and Trower, W.P., Particle Physics in the Cosmos, Readings from Scientific American, W.H. Freeman, San Francisco 1989.
Carrigan, R.A. and Trower, W.P., Particles and Forces: At the Heart of the Matter, Readings from Scientific American, W.H. Freeman, San Francisco, 1990.
Davies, P.C.W., The Edge of Infinity, Dent, London, 1981.
Davies, P.C.W. and Brown, J., Superstrings: a theory of everything, Cambridge UP, Cambridge, 1988.
Ferris, T., Coming of Age in the Milky Way, The Bodley Head, London, 1988.
Gribbin, J. and Rees, M.J., Cosmic Coincidences: Dark Matter, Mankind, and Anthropic Cosmology, Bantam, NY, 1989.
Hawking, S.W., A Brief History of Time, Bantam, NY, 1988.
Kaufmann, W. 1977. The Cosmic Frontiers of General Relativity, Little Brown, Boston, 1977.
Krauss, L. The Fifth Essence: the search for dark matter in the universe, Basic Books, NY, 1989.
Lederman, L. and Schramm, D.N., From Quarks to the Cosmos: tools of discovery, Scientific American Library, W.H. Freeman, 1989.
Pagels, H., Perfect Symmetry, M. Joseph, London, 1985.
Silk, J., The Big Bang (2nd ed.), W.H. Freeman, San Francisco, 1988.
Weinberg, S., Dreams of an Ultimate Theory, Basic Books NY, 1993.
Zee, A., Fearful Symmetry, Macmillan, NY, 1986.

Further Reading

Intermediate

Barrow, J.D., The World Within the World, Oxford UP, Oxford, 1988.
Berendzen, R., Hart, R. and Seeley, D., Man Discovers the Galaxies, Science History Publications, NY 1976.
Close, F., The Cosmic Onion, Heinemann, London, 1983.
Cornell, J., ed. Bubbles, Voids and Bumps in the New Cosmology, Cambridge UP, Cambridge, 1989.
Davies, P.C.W., ed. The New Physics, Cambridge UP, Cambridge, 1989.
Drees, W., Beyond the Big Bang: Quantum Cosmology and God, Open Court, La Salle, 1990.
Longair, M., The Origins of the Universe, Cambridge UP, Cambridge, 1990.
Luminet, J-P., Black Holes, Cambridge UP, Cambridge, 1993.
Peat, F.D., Superstrings and the Search for a Theory of Everything, Contemporary Books, Chicago, 1988.
Shu, F. The Physical Universe, Mill Valley, Calif., University Science Books, 1982.
Tayler, R.J., Hidden Matter, Ellis Horwood, Chichester, 1991.
Weinberg, S., The First Three Minutes, Basic Books, NY, 1977.

Technical

Barrow, J.D. and Tipler, F.J., The Anthropic Cosmological Principle, Oxford UP, Oxford and NY, 1986.
Kolb, E. and Turner, M.S., The Early Universe, Addison Wesley, Redwood City Calif., 1990.
Linde A., Particle Physics and Inflationary Cosmology, Harwood, NY, 1990.
Narlikar, J.V., Introduction to Cosmology (2nd edition), Cambridge UP, Cambridge, 1993.
Padmanabhan, T., Structure Formation in the Universe, Cambridge UP, Cambridge, 1993.
Peebles, P.J., Principles of Physical Cosmology, Princeton UP, Princeton NJ, 1993.
Tryon, E.P., Cosmic Inflation, in The Encyclopedia of Physical Science and Technology, vol. 3, pp. 709–43, Academic Press, NY, 1987.
Vilenkin, A. and Shellard, E.P.S., Cosmic Strings and Other Topological Defects, Cambridge UP, Cambridge, 1993.

Index

Index

Index

Index

Elementary particles *(continued)*
principle, 58–59; weakening of interactions between, 85; what are ultimate?, 217–20; why are there?, 215–17

Eliot, T.S., 44

Elliptical galaxies, 133–37, *134–36*, 138, 139, 142, 143, 235

Ellis, George, 41, 44

End of the world, 224–27

Energy: conservation, 65, 67–68; density, 18–19, 112–13; mass formula, 33; release during phase transition, 109–10; and strength of fundamental forces, 79, 81–83, 85; *see also* Temperature

Entropy, 112, 148, 172–73; defined, 19–20, 234; fluctuations in primordial, 129, 131; per baryon, 235; production, 101; specific, 19–20

Escape velocity, 47, 48, 235; critical for expansion of universe, 188–90, 191

Eschatology, 224, 226

Euclidian geometry, 59, 112–13

eV, *see* Electron volt

Event horizon, 52–53, 54–55, 234, 235; particle production at, 68

Exclusion principle, 145, 234, 238

Expanding universe, 6–7, 9, 10–11, 30, 32, 167, 204; and black body radiation wavelengths, 18–19; density of matter in, 105, 110–11; discovery of, 188; *see also* Expansion rate

Expansion rate, 158, 161–63; 175, 189, 206, 207; critical, 188–90, 191, 206, 230; vs. density, 189; and gravity, 23; of infinite vs. finite universe, *11;* terminal, 161–62

Extraterrestrial life, 205–6, 231

Fairbank, William, 84

Fermi, Enrico, 87

Fermion, 220, 236, 237, 238, 240; defined, 235

First condensations, 129–30

Flat universe, 235

Fornax, 144

Friction, 172; and early universe, 161–62, 192

Friedman, Alexander, 6, 8–9, 30, 32, 175, 194, 235

Friedman universe, 110, 176, 179

Frost, Robert, 224

Fundamental forces of nature, 73–75; breaking of symmetry between, 222–23; in early universe, 85–87, *86;* search for unified theory of, 75–76; unification of and energy levels, 81–82, *82; see also* Electromagnetic force or interaction; Grand unification theories (GUTs); Gravity; Strong force or interation; Weak force or interaction

Future: and end of world, 224–27; finite vs. infinite, 188–90; predictions of, 223–24

Galactic "bulge," 132

Galactic spheroids, 137–38, 239

Galactic year, 132

Galaxies, 21–22, 23, 24, 25; density of vs. velocity of expansion, 189; distances of, 8, 103–4; distribution of, 104–9, *106–7,* 158; elliptical, 133–37, *134–36;* evolution of, 212; formation of, 25, 28, 116–18, 119–23, 129–32, 139–44; and gravitational instability, 236; mergers of, 142–44, *143;* minimum apparent size, 41; nuclei of active, 49–50, 55; number of, 25; origin of, 102–3, 110, 230–31; radiation, 16; rate of recession of, 6, 8, *9,* 10; redshifts of, 7–8, 12; rotation curves, 126, *127;* spiral, 132–37, *134–36;* types of, 132–38; *see also* Clusters;

254

Index

Index

Index

257

Index

Nonlinearity of Einstein's equations, 177–78
Nonsimultaneous creation, 42–43, *43;* see also Continuous creation theory
"Normal" or Gaussian distribution, 214, 238
Novikov, Igor, 18
Nuclear force, *see* Strong nuclear force or interaction
Nuclear reactions, in stars, 147–50
Nucleons, 83, 238

Observable universe, 26–27, 167
Occam's razor, 238
Olbers, Heinrich, 14
Olbers' paradox, 14–15, 238
Omphalos (Gosse), 4
Open universe, 14–15, 188–90; defined, 238; survival in decaying, 226; time in, 187; *see also* Universe, finite vs. infinite
Orion belt, 151–52
Oxygen, 100, 150, 205, 236

Pancakes, 117–18, 119, 120–22, 128, 130, 238; fragments, 131
Parity, 213
Parker, Eugene, 202
Parker limit, 202
Parsec, 238
Particle-antiparticle symmetry, 213; *see also* Matter, imbalance between antimatter and
Past light cone, 168, *168*
Pauli exclusion principle, *see* Exclusion principle
Penrose, Roger, 38, 55, 234
Penzias, Arno, 15, 16, 18, 41
Perforated universe, 36–37
Phases, 190–91

Phase transition, 109–10, 111, 202–3, 238; and inflation, 191, *192,* 194–95
Phosphorus, 205
Photinos, 220, 238
Photon(s), 11–12, 25, 48, 77, 220, 233, 235; baryon ration, 112, 129, 148, 173, 233; in black body radiation field, 18; as boson, 234; defined, 238; energy of and expanding universe, 19; ionization of hydrogen by, 115; microwave, 41; ratio of protons to, 94–95, 97, 101; rest-mass of, 121; and virtual electron-positron pairs, 80
Physical Society of London, 13
Physics, laws of, 209–11; existence of, 213–15; *see also* Nature, laws of
Pi meson or pion, 93, 221, 233, 237, 238
Planck, Max, 62
Planck's constant, 58, 59, 62, 76, 238, 239, 241
Planck spectrum, 234
Planck temperature, 213–14
Planck time or instant (t_p), 62–63, 67–68, 70, 86, 170, 190, 231, 238–39; horizon size at, 108, 110; *see also* Universe, first 10^{-43} second
Planets: atmospheres, 47; formation of, 28, 154
Plasma, 25
Polyakov, Alexander, 197
Positrons, 77, 89, 218, 239; virtual, 77, 80
Pressure effects, 119; absence of, 116; and density fluctuations, 113–14; and infinite density, 33–34
"Primeval atom," 194
Proper time, 186, 187, 239
Protogalactic phase, 144
Protogalaxy, 239
Proton(s), *x,* 79, 83, 84, 88, 89, 90, 92, 96, 217, 218, 224, 233, 235; in black body radiation field, 18; creation of, 94; decay, 92–94; decay, magnetic monopoles and, 198–99; decay, and super particles, 221; defined, 239; fusion of, with neutrons, 99–101; inter-

Index

Index

Index

Index

Vacuum, 76–77, 241; energy, 194–95; fluctuation, 241; polarization, 77, *78*, 81, 241; quantum, 64–67

Vaucouleurs, Gerard de, 137

Velocity of light, *see* Light, velocity of

Virgo supercluster, 21, 22, 23, 24, 147

Virtual particle pairs, 64–66, *64*, 76, 77–78, 80, 241; and black holes, 68–69, *69;* creation of real, detectable particles from, 67–68; effects of, 65, 66–67

Virtual quark-antiquark pairs, 80–81

Viscosity, 162–63, 237, 241

Voids, 241; evolution of, 118; formation of, 116–17

$\underline{V}$ rishon, 218, 219, 239

$\overline{V}$ rishon, 218

Wallace, Alfred, 228

"Walls" of energy, 203, 241

Wavelength, 241

Wave-particle duality, 57–58

W boson or particle, 81, 96, 220, 234; defined, 241; in magnetic monopoles, 197–98

Weak force or interaction, 74, 75, 85, 88, 109, 162; defined, 241; in early universe, 86–87, 88; and energy levels, 79, 81, *82;* grand unification of, 190, 219; maintenance of symmetry

with electromagnetic force, 96; and neutrinos, 98; particles controlled by, 87; unification of with strong force, 111

Wheeler, John, 48

"Whimper" singularities, 44–46, 241

White dwarf, 241

White holes, 42–44, 241

White noise, 241

Wilson, Robert, 15, 16, 18, 41

Winos, 220, 241

X bosons or particles, 88–89, 92–95, 111–12, 234; and creation of matter, 94–95; decay, 94, 95; defined, 241–42; and final singularity, 225; in magnetic monopoles, 197–98; and proton decay, 92–94; $\overline{X}$ decay rate, 94–95

X-rays or radiation, 14, 119, 126; defined, 242; and detection of black holes, 48–49

Z bosons, 96, 220, 234; defined, 242; and magnetic monopoles, 197–98

"Zero point" motion, 66

Zinos, 220, 242